CHEMICAL ENGINEERING PROCESS ANALYSIS

Chemical Engineering Process Analysis

A. M. MEARNS

Department of Chemical Engineering
The University of Newcastle upon Tyne

OLIVER & BOYD · EDINBURGH

OLIVER AND BOYD
Croythorn House
23 Ravelston Terrace
Edinburgh EH4 3TJ
A Division of Longman Group Limited

ISBN 0 05 001886 8
First published 1973
© 1973 A. M. Mearns
All Rights Reserved

Set in 10/12 Times New Roman 327 series
and printed in Great Britain by
Bell and Bain Ltd, Glasgow

CONTENTS

PREFACE   vii

NOMENCLATURE   ix

1 THERMODYNAMIC FEASIBILITY OF A CHEMICAL REACTION
   1.1   Introduction   1
   1.2   Criterion of Equilibrium   1
   1.3   Free Energy Change and Chemical Reaction   4
   1.4   The Equilibrium Constant and its Relation to Free
         Energy Change for a Reaction   5
   1.5   Standard Free Energy Data   9
   1.6   Equilibrium Composition   17

2 KINETICS AND REACTOR DESIGN
   2.1   Introduction   31
   2.2   Rate Equations for Reactions in Batch Systems   31
   2.3   Design Equations for Flow Reactors   39

3 REVERSIBLE EXOTHERMIC REACTIONS
   3.1   Introduction   57
   3.2   Reversible First Order Reactions   57
   3.3   Synthesis of Ammonia   72

4 STEAM REFORMING OF HYDROCARBONS
   4.1   Introduction to Steam Reforming   94
   4.2   Further Treatment of Reformed Gases for Synthesis Gas
         Preparation   113

5 HYDROCARBON CRACKING AND OLEFINE MANUFACTURE
   5.1   Introduction   121
   5.2   Thermodynamics of the Cracking Process   121
   5.3   Kinetics of Hydrocarbon Cracking Reactions   124
   5.4   Conditions of Operation of Pyrolysis Reactors   133
   5.5   Kinetic Models for Hydrocarbon Pyrolysis Reactions   137
   5.6   Use of Kinetic Models in the Design of Pyrolysis Re-
         actors   139

6 MULTI-STEP REACTION PROCESSES
   6.1   Introduction   153
   6.2   Acetylene Production from Hydrocarbons   154
   6.3   Parallel-consecutive Reactions   168

7 EXOTHERMIC CATALYTIC REACTIONS—PHTHALIC ANHYDRIDE
  SYNTHESIS
  7.1   Introduction   182
  7.2   Phthalic Anhydride Synthesis from Naphthalene   183

8 HOMOGENEOUS CATALYSIS BY CO-ORDINATION COMPOUNDS OF
  TRANSITION METALS—WACKER PROCESS FOR ACETALDEHYDE
  SYNTHESIS
  8.1   Introduction   207
  8.2   Homogeneous Catalysis by Co-ordination Compounds of
        Transition Metals   207
  8.3   The Wacker Process   216
  8.4   Mass Transfer Effects in Gas–Liquid Reactions   225

APPENDIX 1   Standard Heats, Free Energies and Entropies of For-
             mation of some Common Compounds at 298°K   233
             Molar Heat Capacity of Gases in Ideal Gas State (298–1500°K)   236

APPENDIX 2   Free Energy and Function for Ideal Gas State   237
             Enthalpy above 0°K for Ideal Gas State   238

INDEX   239

PREFACE

The type of chemical reactor used for a particular process, and conditions of operation with respect to such variables as temperature, pressure and flow rates of reactants, depend largely on the kinetics of the reactions involved, and also on the degree to which equilibrium conditions are approached within the reactor. Many books deal with the general principles of chemical engineering kinetics and reactor design, and the scope and utility of equilibrium thermodynamics are considered in all books dealing with the subject of chemical engineering thermodynamics. However, applications of these principles to an analysis of reactor design and conditions of operation for specific chemical processes, and discussion of the necessary simplifications and approximations when dealing with real systems, are at present to be found only in original papers. In this book an attempt is made to gather together information relating to selected processes, to illustrate the manner in which fundamental ideas of kinetics and thermodynamics are used to predict process operating conditions, as well as any modifications which must be made to these when analysing complex systems, and to suggest the most appropriate form of reactor for the process being considered.

Chapters 1 and 2 present the necessary thermodynamic and kinetic concepts and equations required in the remainder of the book. These chapters constitute summaries only, and prior knowledge of these subjects is desirable for a full understanding of the presented material.

In Chapter 3 the features of reversible exothermic reactions, which include some industrial processes of major importance, are analysed. The synthesis of ammonia has been chosen for discussion since the characteristics of this process are well documented. Equilibrium considerations are of prime importance in this class of reactions, as they are in the hydrocarbon steam reforming process, which is discussed in Chapter 4. The latter process exemplifies the use of simultaneous reaction equilibria equations for the elucidation of product compositions.

Aside from the fact that thermodynamics predicts the approximate temperature threshold for successful reaction, the design and operation of reactors for the pyrolysis of hydrocarbons are largely determined by reaction kinetic considerations. Although the order of reaction can, to a first approximation, be considered simple, one has to resort to semi-empirical ideas to obtain satisfactory guide-lines for deciding on conditions of operation. Hydrocarbon cracking processes constitute the subject matter of Chapter 5.

Features of processes exhibiting more complex kinetics, in which operating conditions can be fairly successfully optimised, are discussed in Chapter 6. Acetylene formation from hydrocarbons is used as an illustration of a consecutive reaction and the chlorination of benzene is discussed as an example of a consecutive-competitive reaction.

The dissipation of heat is a major problem in many exothermic heterogeneous gas–solid catalytic reactions, and the interaction of these heat effects on reaction rates must be accounted for by a simultaneous consideration of mass and energy balance equations. Approaches to the problem of the design and operation of reactors for this type of process are covered in Chapter 7 in which naphthalene oxidation is used as an illustration.

Liquid phase oxidation reactions, homogeneously catalysed by transition metal ions, are of growing importance in the chemical industry. The general principles of homogeneous catalysis and some aspects of transfer of gas into liquids, which are of particular relevance in this process, are discussed in the final chapter. The Wacker process for the synthesis of acetaldehyde by oxidation of ethylene is chosen as an example of this class of reaction, and the mechanism of reaction and conditions of operation are analysed in some detail.

NOMENCLATURE

A_1, A_2	Arrhenius frequency factors
A_c	Cross-sectional area of reactor
A_w	Area of reactor wall/unit length
C	Concentration of species
C_0	Initial concentration of species
C_{A_i}	Concentration of species A at phase inter-phase
C_p	Heat capacity
D_A	Diffusion coefficient of species A
D_e, D_{e_r}, D_{e_L}	Effective diffusivity, effective diffusivity in radial direction, and in axial direction
D_p	Bubble diameter
E_A	Expansion factor-fractional change in volume of system between no conversion and complete conversion
E_1, E_2	Energies of activation
F	Molar flow rate of species
F_0	Initial molar flow rate of species
G	Superficial mass velocity/unit area
$G, G^\circ, \Delta G, \Delta G^\circ, G_0{}^\circ$	Free energy, standard free energy, free energy change, standard free energy change, free energy at 0°K.
$H, H^\circ, \Delta H, \Delta H^\circ, H_0{}^\circ$	Enthalpy, standard enthalpy, enthalpy change, standard enthalpy change, enthalpy at 0°K
$K, K_f, K_f{}', K_p$	Equilibrium constant, equilibrium constant in terms of fugacities, in terms of fugacities for an ideal gas mixture, in terms of partial pressures
K_γ	Fugacity coefficient equilibrium constant
K_{app}	Apparent equilibrium constant
$L, \Delta L$	Length, increment of length
N_A	Average rate of mass transfer of species A/unit area

P	Total pressure
R, R_{av}	Rate of reaction, average rate of reaction
$S, S^{\circ}, \Delta S, \Delta S^{\circ}$	Entropy, standard entropy, entropy change, standard entropy change
T, T_m, T_w, T_c, T_R	Temperature, mean temperature, wall temperature, critical temperature, reduced temperature
U	Overall heat transfer coefficient
ΔU	Internal energy change
$V, V_C, V_D, \bar{V}$	Volume of reactor, continuous phase volume, gas hold-up volume, partial molar volume
W	Work done on system
X, Y	Conversion of species
X_e, X_{me}	Conversion of species at equilibrium, mean conversion
$Z, \Delta Z$	Length, increment of length
a	Activity, interfacial area
d_t, d_p	Tube diameter, particle diameter
$f, f^{\circ*}, f^{\circ}, f'$	Fugacity, fugacity in undefined standard state, fugacity in standard state of 1 atmosphere, fugacity of pure gas at temperature and pressure of mixture
h	Fluid bed height, static gas hold-up
h', h_w	Dynamic gas hold-up, wall heat transfer coefficient
$k, k_L, k_L{}^*$	Rate constant, liquid mass transfer coefficient, liquid mass transfer coefficient enhanced by reaction
k_e, k_{er}, k_{e_L}	Effective conductivity, effective conductivity in radial and axial directions
n_i, n_1, n_T	Number of moles of species, number of moles of inerts, total number of moles
n_{i_o}	Number of moles initially
n	Number of reactors
p_i, p_c, p_R	Partial pressure of species, critical pressure, reduced pressure
$q, \Delta q, \Delta q_c, \Delta q_L$	Heat change in system, heat supplied per reactor tube, heat of cracking, latent heat change

r, r_0	Radial length, radius of reactor
$(-r_A), (-r_{c_A})$	Rate of reaction (disappearance) of species A, rate of reaction in catalytic reactor
$t, \bar{t}$	Time, mean holding time in stirred tank reactor
u, u_{SG}, u_{SL}, u_T	Superficial velocity, superficial gas velocity, superficial liquid velocity, terminal bubble velocity
v	Volumetric flow rate
x_i	Mole fraction of species
α	Order of reaction
α_i	Rate constant ratio
β	Order of reaction, concentration ratio, C_C/C_{A_0}
γ	Concentration ratio, C_A/C_{A_0}
γ, γ'	Fugacity coefficient, fugacity coefficient at total pressure P
δ	Expansion factor, film thickness
ε	Voidage of catalyst bed
$\mu, \mu^{\circ*}, \mu^{\circ}$	Chemical potential, chemical potential in an undefined standard state, chemical potential of pure gas at 1 atmosphere pressure
μ_L	Liquid viscosity
v_i	Stoichiometric coefficient
ρ, ρ_B, ρ_L	Density, bulk density of catalyst, liquid density
σ_L	Liquid surface tension
τ	Space time
ϕ	Mass transfer coefficient ratio, k_L^*/k_L
ψ	Shape factor
Pe	Peclet number
Re	Reynolds number
Sc	Schmidt number
Sh	Sherwood number
$S.T.Y$	Space time yield
$S.V.$	Space velocity

Subscripts

I	Irreversible
R	Reversible

T	Temperature
0	Absolute zero of temperature
av	Average
f	Formation
i	Species identifier
r	Reaction
sys	System
surrs	Surroundings

Note regarding system of units:

In general SI units have been employed throughout. Temperature (Kelvin) has been assigned the symbol K in the diagrams but written as °K in the text to avoid confusion with the Equilibrium Constant symbol, K.

1

THERMODYNAMIC FEASIBILITY OF A CHEMICAL REACTION

1.1 Introduction

In devising a chemical process two crucial questions must be answered.

What are the conditions of temperature and pressure necessary for the reaction to proceed and will these be attainable under plant operating conditions?

What will be the theoretical product yield at the chosen reaction conditions?

Fortunately, it is possible to predict the answers through the application of the concepts of equilibrium thermodynamics. It must be appreciated, however, that the predicted product yields may not be attained in practice, because the reaction rates may be too slow for equilibrium concentrations to be reached in a realistic time spent in the reactor. Thermodynamics is of no help in estimating the rates of chemical reactions.

Methods for predicting reaction feasibility and for the calculation of equilibrium yields will be outlined in this chapter. It does not aim to give a comprehensive discussion of the application of thermodynamics to chemical processes, but presents some of the criteria for successful analysis of processes considered in later chapters. For more complete thermodynamic coverage see, for example, Smith and Van Ness (1959), Dodge (1944), or Hougen, Watson, and Ragatz (1959). The rates of chemical reactions will be considered in Chapter 2.

1.2 Criterion of Equilibrium

Equilibrium may be thought of as a state where all forces are balanced and the net driving force for any change is zero. A process where the driving force has approached zero is termed reversible, and, if the

1

process is carried out under these conditions, the maximum possible amount of work will be obtained. If carried out under the influence of a finite driving force, such that the amount of work produced is less than maximum, the process is said to be irreversible or spontaneous. It follows that

$$W_R - W_I > 0, \qquad (1.1)$$

where W_R is the work obtained if the process is operated reversibly and W_I is the work obtained under any other condition of operation.

By the First Law of Thermodynamics, for a change in the system from state 1 to state 2,

$$\Delta U = q - W, \qquad (1.2)$$

where ΔU is the change in internal energy of the system and q is the heat change occurring in the system. Since ΔU is a property of the system the change in internal energy is the same irrespective of the path followed in going from state 1 to state 2. Hence, by application of equation (1.2) to the process carried out both reversibly and irreversibly, it follows that

$$q_R - q_I > 0, \qquad (1.3)$$

i.e., the heat effect when the process occurs reversibly is greater than for an irreversible change, since $W_R > W_I$.

The above discussion has been concerned solely with the system, but when a heat change occurs in the system an equal and opposite change must occur in the surroundings. The heat change which occurs in the surroundings will occur reversibly because the surroundings are infinite in nature compared with any system and the temperature of the surroundings will not change.

If we consider a system undergoing an irreversible, isothermal change from state 1 to state 2, in which the system and surroundings are at the same temperature, thus eliminating other sources of irreversibility, the change in entropy of the system, written as ΔS_{sys} is given by

$$\Delta S_{sys} = q_R/T, \qquad (1.4)$$

which is, by definition, the change in entropy for a change at constant temperature. Because the heat change, q_I, which takes place in the surroundings is reversible the related entropy change will be given by $-q_I/T$ (the negative sign indicates that it will be in the opposite sense to the change occurring in the system).

The total entropy change in the system plus surroundings is therefore given by

$$\Delta S = \Delta S_{sys} + \Delta S_{surrs} = q_R/T - q_I/T. \qquad (1.5)$$

From equation (1.3), for an irreversible change in the system, ΔS will be greater than zero but when the change from state 1 to state 2 occurs reversibly it follows that $\Delta S = 0$.

One criterion for equilibrium is thus that for any change of state the sum of the changes of the entropy of system plus surroundings shall be zero.

In general, it is more convenient to refer to the system alone rather than to the system plus surroundings. For a process which occurs reversibly the heat change in the system, q_R, is related, by equation (1.4), to the entropy change and since, by equation (1.3), the heat change for a reversible process is greater than for an irreversible, it follows that

$$T\Delta S > q_I, \qquad (1.6)$$

where ΔS refers to the system.

This equation can be generalised to

$$T\Delta S - q \geqq 0, \qquad (1.7)$$

the inequality sign applying to an irreversible, and the equality sign to a reversible, isothermal process. Further, in the case when pressure remains constant during the process, $q = \Delta H$, where ΔH is the enthalpy change occurring, and hence by substitution in (1.7) we obtain

$$T\Delta S - \Delta H \geqq 0. \qquad (1.8)$$

A new term ΔG, which is called the free energy change for the system, is introduced to define this difference—

$$\Delta G = \Delta H - T\Delta S. \qquad (1.9)$$

Hence, equation (1.8) may be rewritten as

$$G \leqq 0. \qquad (1.10)$$

Equation (1.10) indicates that if a change of state can occur, at constant temperature and pressure, which will lead to a decrease in free energy of the system, such a change will be irreversible and will occur spontaneously, i.e. will not be an equilibrium process. If, however, the free energy change involved is zero, no change in state can occur and the system is at equilibrium.

1.3 Free Energy Change and Chemical Reaction

When the system under consideration consists of chemical substances capable of reacting together, the feasibility of reaction may be assessed by calculating the free energy change involved in converting reactants to postulated products. If the difference between the free energy content of the products and the reactants, measured in relation to the stoichiometry of the reaction, is negative, i.e. if the reaction can proceed in such a manner that there is a decrease in the free energy, it can be said that the reaction will be spontaneous and formation of products will be favoured. On the other hand, if ΔG for the reaction is positive it can be concluded that, under the conditions of pressure and temperature to which the free energy data relate, the driving force in the direction of product formation is negative and a successful process would not be possible.

To calculate the free energy change for a given chemical reaction, values of the free energy of formation of each compound must be available. An enormous amount of data would have to be amassed in order to calculate free energy changes for reaction under all possible conditions of temperature and pressure and, in practice, free energies of formation are usually tabulated at unit fugacity (which may be taken as equivalent to a pressure of 1 atmosphere) and usually at a temperature of 298°K. The free energy change calculated from free energies of formation at unit fugacity is called the standard free energy change, $\Delta G°$. The convention has been adopted of relating the standard free energies of formation of compounds to those of the most stable form of the elementary substances from which they are formed, the standard free energies of formation of these elementary substances being taken as zero.

Standard free energies of formation of representative compounds at 298°K are presented in Appendix 1.

Example 1.1. What is the free energy change for the reaction

$$CH_4(g) + 2O_2(g) = CO_2(g) + 2H_2O(l),$$

at 298°K and atmospheric pressure?
This problem is easily solved, since the change takes place at atmo-

spheric pressure. Standard state data, which are readily available, may thus be used.

From Appendix 1, at 298°K, $(\Delta G°)_f$ for $CH_4(g) = -50·79$ kJ/mol

$$O_2(g) = 0 \quad \text{kJ/mol}$$

$$CO_2(g) = -394·4 \text{ kJ/mol}$$

$$H_2O(l) = -237·2 \text{ kJ/mol}$$

$$(\Delta G°)_{reaction} = (-394·4) + (2 \times -237·2) - (-50·79)$$

$$= -818·01 \text{ kJ/mol}.$$

Since the calculated free energy change has a high negative value, the oxidation of methane to carbon dioxide and water is highly favoured thermodynamically under these 'standard' conditions. The feasibility of reaction may, however, be very different under other conditions.

1.4 The Equilibrium Constant and its Relation to Free Energy Change for a Reaction

In the previous section, it was shown how the feasibility of a chemical reaction may be assessed in terms of the free energy change for the reaction. But, since free energy data are available in general only for substances in their standard states, the product yields, which are in effect a quantitative measure of the feasibility of reaction, at conditions other than standard, must be deduced using these standard free energy data. It is possible to relate the reacting species concentrations at equilibrium, when the free energy change is zero, to the free energy change when the species are in their standard states, by introducing the concept of chemical potential and by relating the free energy change to the absolute sum of the chemical potentials of the species present when equilibrium is reached.

1.4.1 CHEMICAL POTENTIAL

Consider a system containing n_1 moles of component 1, n_2 moles of 2, etc. The free energy of the system may be written in terms of pressure, temperature and composition.

$$G = f(T, P, n_1, n_2, \ldots). \tag{1.11}$$

Hence

$$dG = (\partial G/\partial T)_{P,n_1,n_2} \ldots \ldots dT + (\partial G/\partial P)_{T,n_1,n_2} \ldots dP$$

$$+ (\partial G/\partial n_1)_{T,P,n_2} \ldots dn_1 + \ldots \quad (1.12)$$

Now

$$G = H - TS = U + PV - TS. \quad (1.13)$$

Hence

$$dG = dU + PdV + VdP - TdS - SdT. \quad (1.14)$$

For a reversible change involving only work of expansion

$$dU = TdS - PdV, \quad (1.15)$$

by application of the First Law of Thermodynamics and equation (1.4).

Combining (1.14) and (1.15), it follows that

$$dG = VdP - SdT. \quad (1.16)$$

Further, under conditions of constant temperature, we see that

$$(\partial G/\partial P)_T = V, \quad (1.17)$$

while at constant pressure

$$(\partial G/\partial T)_P = -S. \quad (1.18)$$

Substituting equations (1.17) and (1.18) in (1.12) we obtain

$$dG = -S\,dT + V\,dP + \mu_1 dn_1 + \mu_2 dn_2 \ldots, \quad (1.19)$$

where

$$\mu_1 = (\partial G/\partial n_1)_{T,P,n_2} \ldots$$

is the chemical potential or partial molar free energy of component 1 in the mixture.

At constant temperature and pressure it follows that

$$dG = \mu_1 dn_1 + \mu_2 dn_2 \ldots, \quad (1.20)$$

and on integrating

$$G = \mu_1 n_1 + \mu_2 n_2 \ldots \quad (1.21)$$

1.4.2 EFFECT OF PRESSURE ON CHEMICAL POTENTIAL

If temperature and all composition quantities $n_1, n_2 \ldots$ except n_i remain constant in equation (1.12) then

$$dG = VdP + \mu_i dn_i. \quad (1.22)$$

$$(\partial\mu_i/\partial P)_{T,n_1,n_2}\ldots = (\partial V/\partial n_i)_{T,P,n_1}\ldots, \text{ by Euler's theorem,} \quad (1.23)$$

where

$$(\partial V/dn_i)_{T,P,n_1}\ldots = \overline{V}_i$$

is the partial molar volume of component i.

Hence

$$d\mu_i = \overline{V}_i \cdot dP, \quad (1.24)$$

which on integration gives

$$\mu_{i_{P_2}}-\mu_{i_{P_1}} = \int_{P_1}^{P_2} \overline{V}_i \cdot dP. \quad (1.25)$$

When equation (1.25) is applied to a mixture of ideal gases where $\overline{V}_i = (RT/p_i)$, we obtain the expression

$$\mu_i = \mu_i^{\circ*} + RT \ln (p_i/p_i^{\circ*}), \quad (1.26)$$

where $\mu_i^{\circ*}$ represents the chemical potential of component i in a standard state. If the standard state is chosen as pure gas i at a pressure of 1 atm., equation (1.26) becomes

$$\mu_i = \mu_i^{\circ} + RT \ln p_i, \quad (1.27)$$

where μ_i° is the chemical potential in the standard state.

Likewise when applied to a mixture of real gases, where the fugacity of a component may be substituted for partial pressure, we can write

$$\mu_i = \mu_i^{\circ*} + RT \ln (f_i/f_i^{\circ*}), \quad (1.28)$$

and when the standard state is defined as that where the fugacity of the pure gas is unity,

$$\mu_i = \mu_i^{\circ} + RT \ln f_i. \quad (1.29)$$

In general, equation (1.28) may be written as

$$\mu_i = \mu_i^{\circ*} + RT \ln a_i, \quad (1.30)$$

where $a_i = (f_i/f_i^{\circ*})$ is called the activity of component i, and $\mu_i^{\circ*}$ is the chemical potential of i in its standard state of unit activity.

1.4.3 THE EQUILIBRIUM CONSTANT

Suppose that the chemical reaction

$$a\text{A}+b\text{B} \rightleftharpoons c\text{C}+d\text{D}$$

proceeds at constant temperature and pressure to an extent such that

c moles of C and d moles of D are formed from a moles of A and b moles of B. By equation (1.21), the free energy change will be given by

$$\Delta G = (c\mu_C + d\mu_D) - (a\mu_A + b\mu_B). \qquad (1.31)$$

The chemical potential of a component in a mixture has been given in a general form by equation (1.30), and if this is substituted in (1.31), for each component of the reacting mixture we obtain

$$\Delta G = [c(\mu_C{}^\circ + RT \ln a_C) + d(\mu_D{}^\circ + RT \ln a_D)] - [a(\mu_A{}^\circ + RT \ln a_A)$$
$$+ b(\mu_B{}^\circ + RT \ln a_B)]. \qquad (1.32)$$

Equation (1.32) may be rearranged to give

$$\Delta G = \Delta G^\circ + RT \ln \frac{a_C{}^c a_D{}^d}{a_A{}^a a_B{}^b}, \qquad (1.33)$$

where, by application of equation (1.21) to components in their standard states,

$$\Delta G^\circ = (c\mu_C{}^\circ + d\mu_D{}^\circ) - (a\mu_A{}^\circ + b\mu_B{}^\circ).$$

This equation could be used to determine the feasibility of the reaction where the starting activities are known.

As the process continues the activities change until at equilibrium, from equation (1.10), $\Delta G = 0$. In that circumstance equation (1.33) becomes

$$\Delta G^\circ = -RT \ln \frac{a_C{}^c a_D{}^d}{a_A{}^a a_B{}^b}, \qquad (1.34)$$

$$= -RT \ln K,$$

where

$$K = \frac{a_C{}^c a_D{}^d}{a_A{}^a a_B{}^b}.$$

K is the equilibrium constant for the reaction in terms of activities.

Equation (1.34) is perfectly general and will apply for any reaction equilibrium, irrespective of whether the system is homogeneous or heterogeneous. The equilibrium constant depends on the temperature and on the free energy change which accompanies the reaction of the indicated number of moles of each reactant initially present in its standard state of unit activity, and each product present finally in its standard state of unit activity. Provided ΔG° is known at the temperature in question, equation (1.34) may be used, in principle, to determine the equilibrium composition of a reaction mixture at any value of pressure.

It is important to note that the value of $\Delta G°$ is specific for the indicated stoichiometry of the reaction and only has meaning when the temperature is specified and the standard states of each component are defined.

1.5 Standard Free Energy Data

Before going on to consider the calculation of equilibrium compositions and to discuss how these are affected by changes in reaction conditions, methods of evaluating standard free energy changes at any required temperature will be examined.

Standard free energy data available in the literature are in general of two types. In the first (see Appendix 1), free energies of formation of compounds, usually measured at 298°K, are presented, the standard state of unit activity for a gas being that where the fugacity of the pure gas is unity, and for liquids and solids the pure substance at atmospheric pressure. The values quoted are relative to the free energies of formation of elementary substances in their most common states which are assumed to be zero under standard state conditions. The standard free energy of formation of a compound is then defined as the change in free energy when one mole of the compound is formed from its elements with each substance in its standard state at 298°K. In a variant of this first presentation the standard free energy of formation may be calculated from the standard heat of formation of the compound and the standard entropy change by means of equation (1.9), $\Delta G° = \Delta H° - T\Delta S°$ (Appendix 1). Compilations giving values of $\Delta G_f°$, $\Delta H_f°$ and $S_f°$ at 298°K for many compounds include those published by the National Bureau of Standards (1952 and later supplements) and the American Petroleum Institute (1952 and later supplements). If $\Delta G_f°$ values are not also tabulated at other temperatures they must be calculated, using the methods to be presented in the next sub-section.

The second method of presentation of free energy data enables quick calculation of standard free energy changes to be made at any temperature. This method of presentation gives data for the function

$$-\frac{(G_T° - H_0°)}{T}$$

over the temperature range from 0 to 1500°K, together with the standard heat of formation of compounds at absolute zero. To enable calculation of $\Delta H_f{}^\circ$ at any temperature, values for the function $(H_T{}^\circ - H_0{}^\circ)$ are also presented. These functions give values for the difference between the standard free energy of formation of a compound at temperature T and absolute zero, divided by T, and the standard heat of formation at T and absolute zero respectively. According to the convention adopted for this method, the standard free energy and heat of formation of elementary substances at absolute zero are taken as zero. Values of the functions for selected compounds are given in Appendix 2. Values of $\Delta G_f{}^\circ$ and $\Delta H_f{}^\circ$ at temperatures other than those tabulated may be obtained by interpolation.

An alternative method of presentation is to be found in the JANAF Tables (1965). These tables present free energy functions relative to values at 298°K, for inorganic compounds, i.e. values of $-(G^\circ - H_{298}{}^\circ)/T$ are given.

The standard free energy of a compound relative to its elements at any temperature can be obtained from the following equation:

$$\left(\frac{\Delta G_f{}^\circ}{T}\right)_T = \left[\frac{(G_T{}^\circ - H_0{}^\circ)}{T} + \frac{(\Delta H_f{}^\circ)_0}{T}\right]_{\text{compound}}$$
$$- \sum \left[\frac{(G_T{}^\circ - H_0{}^\circ)}{T}\right]_{\text{elements}}. \tag{1.35}$$

In a like manner the standard heat of formation of a compound relative to its elements at any temperature is given by

$$(\Delta H_f{}^\circ)_T = [(H_T{}^\circ - H_0{}^\circ) + (\Delta H_f{}^\circ)_0]_{\text{compound}} - \sum [(H_T{}^\circ - H_0{}^\circ)]_{\text{elements}}. \tag{1.36}$$

By combining equations for $(\Delta G_f{}^\circ)_T$ and $(\Delta H_f{}^\circ)_T$ for all reactants and products in the appropriate stoichiometric amounts, the following equations are obtained for the standard free energy and standard heat of reaction:

$$\left(\frac{\Delta G^\circ}{T}\right)_T = \sum \left[\left(\frac{G_T{}^\circ - H_0{}^\circ}{T}\right) + \frac{(\Delta H_f{}^\circ)_0}{T}\right]_{\text{products}}$$
$$- \sum \left[\frac{(G_T{}^\circ - H_0{}^\circ)}{T} + \frac{(\Delta H_f{}^\circ)_0}{T}\right]_{\text{reactants}}. \tag{1.37}$$

$$(\Delta H^\circ)_T = \sum \left[(H_T^\circ - H_0^\circ) + (\Delta H_f^\circ)_0 \right]_{\text{products}}$$
$$- \sum \left[(H_T^\circ - H_0^\circ) + (\Delta H_f^\circ)_0 \right]_{\text{reactants}}. \quad (1.38)$$

By writing a combination equation for products and reactants the terms involving free energies and heats of formation of the elementary substances have cancelled.

It will be noticed that the term H_0° appears in the free energy function rather than the expected G_0° and the term $(\Delta H_f^\circ)_0$ is used rather than $(\Delta G_f^\circ)_0$. They are a consequence of the Third Law of Thermodynamics. The Third Law is based on the experimental observation that as absolute zero is approached the standard free energy and heat of formation of a compound tend to the same value and at absolute zero it is postulated that they are identical. Hence $(\Delta H_f^\circ)_0$ may be written for $(\Delta G_f^\circ)_0$ and H_0° for G_0°. (See section 1.5.1 for further discussion of the utility of the Third Law.)

Example 1.2. Calculate the equilibrium constant for dehydrogenation of propane to propylene at 800°K. Refer to Appendix 2 for data.

$$C_3H_8(g) = C_3H_6(g) + H_2(g).$$

Equation (1.37) is used for this calculation:

$$\left(\frac{\Delta G^\circ}{T} \right)_T = \sum \left[\frac{(G_T^\circ - H_0^\circ)}{T} + \frac{(\Delta H_f^\circ)_0}{T} \right]_{\text{products}}$$
$$- \sum \left[\frac{(G_T^\circ - H_0^\circ)}{T} + \frac{(\Delta H_f^\circ)_0}{T} \right]_{\text{reactants}}.$$

Using data from Appendix 2,

$$\frac{\Delta G^\circ}{800} = \left[-280 \cdot 49 + \frac{35430}{800} - 130 \cdot 48 \right] - \left[-287 \cdot 61 - \frac{81510}{800} \right].$$

$$\frac{\Delta G^\circ}{800} = 22 \cdot 82 \text{ J/mol}^\circ\text{K},$$

or

$$\Delta G^\circ_{800 \, ^\circ\text{K}} = 18256 \text{ J/mol}.$$

By application of equation (1.34) the equilibrium constant may be calculated.

$$\log K = -\frac{\Delta G^\circ}{2 \cdot 303 \times RT} = \frac{-18256}{2 \cdot 303 \times 8 \cdot 314 \times 800} = -1 \cdot 192,$$

and

$$K = 0.064.$$

1.5.1 Effect of Temperature on the Standard Free Energy Change for a Reaction and on the Equilibrium Constant

Literature data giving values of $(\Delta G_f^\circ)_{298^\circ K}$ are not of general applicability and it is necessary that the temperature dependence be considered to allow free energies to be calculated at temperatures other than 298°K. Such values can be obtained by application of the van't Hoff equation.

From equation (1.9),

$$\Delta G^\circ = \Delta H^\circ - T \Delta S^\circ,$$

and by substitution of equation (1.34) in (1.9) we obtain

$$-RT \ln K = \Delta H^\circ + T \left(\frac{\partial(-RT \ln K)}{\partial T} \right)_P, \qquad (1.39)$$

where ΔS° has been replaced by

$$-\left(\frac{\partial(\Delta G)}{\partial T} \right)_P$$

according to equation (1.18).

Differentiation of the last term in (1.39) and rearrangement leads to

$$\left(\frac{\partial \ln K}{\partial T} \right)_P = \frac{\Delta H^\circ}{RT^2}. \qquad (1.40)$$

This equation is known as the van't Hoff equation.

If ΔH° is constant over the temperature range of interest, (1.40) may be integrated between T_1 and T_2 to give

$$\ln \frac{K_1}{K_2} = -\frac{\Delta H^\circ}{R} \left(\frac{1}{T_1} - \frac{1}{T_2} \right). \qquad (1.41)$$

This equation may be represented graphically by plotting $\log K$ versus $1/T$ (see Fig. 1.1).

It should be noted that if the reaction is endothermic, i.e. if ΔH° is positive, the equilibrium constant increases with increase in temperature but if exothermic K decreases with temperature. As will be seen in Chapter 3 this fact is of great importance when deciding on reaction conditions for reactions which are of the type described as reversible and exothermic.

A more common case is that where ΔH° is not constant over the whole range of temperature under consideration. It is desirable in such

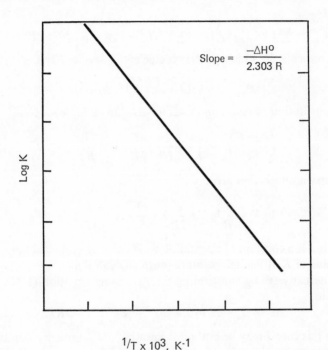

Slope $= \dfrac{-\Delta H^\circ}{2.303\ R}$

Log K

$1/T \times 10^3,\ K^{-1}$

FIG. 1.1. Variation of equilibrium constant with temperature

cases to present ΔH° as a power series expression in T of the form

$$\Delta H^\circ = aT + \tfrac{1}{2}bT^2 \ldots + I_E, \tag{1.42}$$

where I_E is a constant of integration which may be evaluated knowing a value of ΔH° at a particular temperature, usually $298°K$. The form of equation (1.42) follows from the manner in which standard molar heat capacities are commonly represented, i.e., $C_p^\circ = \alpha + \beta T + \gamma T^2 \ldots$ where α, β and γ are constants. The standard heat of formation of a compound is then represented by

$$\Delta H_f^{\circ} = \int C_p^{\circ}.dT = \int (\alpha + \beta T + \gamma T^2 \ldots)dT$$

$$= \alpha T + \tfrac{1}{2}\beta T^2 + \tfrac{1}{3}\gamma T^3 \ldots + I_E', \quad (1.43)$$

and hence

$$\Delta H^{\circ} = \int \Delta C_p^{\circ}.dT = aT + \tfrac{1}{2}bT^2 + \tfrac{1}{3}cT^3 \ldots + I_E, \quad (1.44)$$

where

$$\Delta C_p^{\circ} = \sum v_i \alpha + (\sum v_i \beta)T + (\sum v_i \gamma)T^2 + \ldots = a + bT + cT^2 + \ldots$$

v_i represents the stoichiometric coefficient of species i and

$$a = \sum v_i \alpha = \sum (v_i \alpha)_{products} - \sum (v_i \alpha)_{reactants}, \text{ etc.}$$

Substitution of equation (1.42) for ΔH° in (1.40) leads to

$$\left(\frac{\partial \ln K}{\partial T} \right)_P = \frac{a}{RT} + \frac{b}{2R} + \frac{cT}{3R} \ldots + \frac{I_E}{RT^2}, \quad (1.45)$$

which on integration gives

$$\ln K = \frac{a}{R} \ln T + \frac{b}{2R} T + \frac{cT^2}{6R} \ldots \frac{-I_E}{RT} + J, \quad (1.46)$$

where J is a constant of integration which may be evaluated knowing a value of K at one temperature (normally 298°K).

Alternatively, by substituting for ΔG° using equation (1.34), we obtain

$$\Delta G^{\circ} = I_E - aT \ln T - \tfrac{1}{2}bT^2 - \tfrac{1}{6}cT^3 \ldots - JRT. \quad (1.47)$$

In this case J may be evaluated knowing ΔG° at one temperature.

In summary, in order to evaluate the equilibrium constant or free energy change in a reaction at any given temperature it is necessary to have available the following data:

(i) heat capacities as a function of temperature for all products and reactants involved in the reaction

(ii) a value of ΔH° for the reaction at one temperature to allow evaluation of I_E.

(iii) a value of ΔG° for the reaction at one temperature to allow evaluation of J or a value of K at one temperature.

The data mentioned in (i) and (ii) are thermal, but those in (iii) are not. Evaluation of ΔG° or K at one temperature directly from thermal data would facilitate calculation. In fact it is possible to use only thermal data by invoking the Third Law of Thermodynamics. The Third Law is based on the observations that as absolute zero is approached $\Delta G - \Delta H \to 0$ and further that

$$\left(\frac{\partial(\Delta H)}{\partial T}\right)_P \to 0 \quad \text{and} \quad \left(\frac{\partial(\Delta G)}{\partial T}\right)_P \to 0$$

as absolute zero is approached. Since

$$\left(\frac{\partial(\Delta G)}{\partial T}\right)_P = -\Delta S$$

it follows that $\Delta S \to 0$ as absolute zero is approached. This has led to the following statement of the Third Law: The entropy change in a reaction between pure crystalline solids approaches zero as the temperature falls to absolute zero. Practically, the absolute entropies for a given compound may be calculated by summing all the entropy changes, including changes of state, involved in bringing the compound from the pure crystalline solid state at absolute zero to the specified state at the desired temperature, the entropy at absolute zero for the pure crystalline solid being taken as zero.

Symbolically this may be written

$$S^{\circ}_{298^\circ K, \text{ compound}} = S^{\circ}_{298^\circ K} - S^{\circ}_{0^\circ K} = \sum \frac{q_L}{T} + \int_{0^\circ K}^{T_1} \frac{dq_{R_1}}{T}$$

$$+ \int_{T_1}^{T_2} \frac{dq_{R_2}}{T} \ldots + \int_{T_n}^{298^\circ K} \frac{dq_{R_{n+1}}}{T} , \qquad (1.48)$$

where q_L signifies a heat effect in a reversible phase change and the integral terms represent the entropy changes in heating successive single phases from initial to final temperature.

The standard entropy change for the reaction, ΔS°, is then given by

$$\Delta S^\circ = \sum (\nu S^\circ)_{\text{products}} - \sum (\nu S^\circ)_{\text{reactants}}, \qquad (1.49)$$

and ΔG° is obtained by application of equation (1.34).

Thus an alternative method which uses thermal data alone is available for evaluation of ΔG° at a particular temperature.

Equation (1.47) may be applied as before.

Example 1.3. Repeat Example 1.2 for the calculation of the equilibrium equation for the dehydrogenation of propane at 800°K, using the following data:

		α	$\beta \times 10^3$	$\gamma \times 10^6$
C_p°	C_3H_8	10·08	239·30	−73·36
(J/mol°K)	C_3H_6	13·61	188·76	−57·49
	H_2	29·07	−0·84	2·01
		$(\Delta G_f^\circ)_{298°K}$	$(\Delta H_f^\circ)_{298°K}$	$(S^\circ)_{298°K}$
		J/mol	J/mol	J/mol°K
	C_3H_8	−23489	−103847	269·91
	C_3H_6	62720	20410	266·94
	H_2	0	0	130·59

Either equation (1.46) or (1.47) may be used. Application of (1.47) allows ΔG° at 800°K to be calculated, after the constants I_E and J have been evaluated using (1.44) and (1.47), by employing the standard heat of formation and free energy of formation data at 298°K.

Formulate equation (1.44) from the heat capacity equations for each component:

$$a = (13·61 + 29·07 - 10·08) = 32·60.$$

$$b = (188·76 - 0·84 - 239·3) \times 10^{-3} = -51·38 \times 10^{-3}.$$

$$c = (-57·49 + 2·01 + 73·36) \times 10^{-6} = 17·88 \times 10^{-6}.$$

Use equation (1.44) to evaluate I_E by inserting ΔH° at 298°K:

$$\Delta H_{298°K}^\circ = 20410 + 103847 = 124257 \text{ J/mol.}$$

Therefore

$$124257 = (32·6 \times 298) + \tfrac{1}{2}(-51·38 \times 10^{-3})(298)^2$$
$$+ \tfrac{1}{3}(17·88 \times 10^{-6})(298)^3 + I_E,$$

$$I_E = 116666 \text{ J/mol.}$$

$\Delta G_{298°K}^\circ$ for the reaction is given by $(62720 + 23489) = 86209$ J/mol and J may be evaluated using equation (1.47):

$$86209 = 116666 - (32·60)(298)(2·303) \log 298$$
$$- \tfrac{1}{2}(-51·38 \times 10^{-3})(298)^2 - \tfrac{1}{6}(17·88 \times 10^{-6})(298)^3$$
$$- J(8·314)(298)$$

$$J = -9·160.$$

At 800°K

$$\Delta G^\circ = 116666 - (32{\cdot}60)(800)(2{\cdot}303) \log 800$$
$$- \tfrac{1}{2}(-51{\cdot}38 \times 10^{-3})(800)^2 - \tfrac{1}{6}(17{\cdot}88 \times 10^{-6})(800)^3$$
$$- (-9{\cdot}160)(1{\cdot}98)(800)$$

$$\Delta G^\circ = 18137 \text{ J/mol.}$$

Hence from equation (1.34)

$$K = 0{\cdot}065$$

$\Delta G^\circ_{298°K}$ may also be evaluated from

$$\Delta G^\circ = \Delta H^\circ - T\Delta S^\circ \quad \text{using (1.49) to evaluate } \Delta S^\circ$$

$$\Delta S^\circ = 266{\cdot}94 + 130{\cdot}59 - 269{\cdot}91 = 127{\cdot}62 \text{ J/mol °K.}$$

Hence

$$\Delta G^\circ = 124257 - (298)(127{\cdot}62)$$

$$= 86226 \text{ J/mol.}$$

This figure may be seen to be very close to that calculated from $(\Delta G_f^\circ)_{298°K}$ data supplied.

1.6 Equilibrium Composition

In the worked example in section 1.3 it was concluded that, because of the negative calculated standard free energy, the formation of product was highly favoured. This was a valid conclusion because the reaction took place under standard state conditions. However, if the feasibility of reaction were required under any reaction conditions other than standard, the value of the calculated ΔG° could be used as a guide only. There are instances of positive calculated ΔG° values, yet the process, as actually carried out, is quite feasible. It must be remembered that for feasible reaction it is ΔG which must have a negative value, not ΔG°. The relation between the equilibrium constant K, related to ΔG°, and composition at equilibrium is considered in this section.

For the chemical reaction

$$a\text{A} + b\text{B} \rightleftharpoons c\text{C} + d\text{D},$$

the equilibrium amounts of substances may be calculated from equation (1.34)

$$K = \frac{a_C{}^c a_D{}^d}{a_A{}^a a_B{}^b}.$$

1.6.1 HOMOGENEOUS SYSTEM—SINGLE GAS PHASE REACTION

The activity of a gas in a gaseous mixture is the ratio of the fugacity of the gas in the mixture, at the temperature and pressure in question, to that in the standard state at the same temperature, where the standard state is that of the pure gas at unit fugacity, i.e. $a = f/f^\circ = f/1 = f$. Hence the equilibrium constant may be written as

$$K_f = \frac{f_C{}^c f_D{}^d}{f_A{}^a f_B{}^b},\qquad(1.50)$$

where f represents the fugacity of a gaseous component in the mixture. By definition, the fugacity of a gas may be given as $f = \gamma p$ where p is the partial pressure of the gas in the mixture and γ is its fugacity coefficient. The equilibrium constant can therefore be written as

$$K_f = \frac{p_C{}^c p_D{}^d}{p_A{}^a p_B{}^b} \cdot \frac{\gamma_C{}^c \gamma_D{}^d}{\gamma_A{}^a \gamma_B{}^b} = K_p \cdot K_\gamma,\qquad(1.51)$$

where K_f represents the equilibrium constant written in terms of fugacities. The equilibrium partial pressures could be calculated from this equation if the fugacity coefficients of the gas in the mixture were known. However, such data are not available since it would require measurement of γ at all possible compositions as well as over a range of temperatures and pressures.

The most usual approach, and one which allows a generally satisfactory solution to be obtained, employs the so-called Lewis-Randall rule which applies if the gas mixture is assumed to be ideal. By this rule the fugacity of a component is related to its mole fraction in the gas by

$$f_i = x_i f_i',\qquad(1.52)$$

where x_i is the mole fraction of component i, f_i' is the fugacity of pure gas i at the temperature and total pressure, P, of the mixture. This assumption is correct only when there is no volume change involved in making the mixture and is liable to greater error as the total pressure increases.

Equation (1.50) may be rewritten assuming the ideal gas mixture assumption holds:

$$K_f' = \frac{x_C^c x_D^d}{x_A^a x_B^b} \cdot \frac{f_C'^c f_D'^d}{f_A'^a f_B'^b}. \tag{1.53}$$

K_f' signifies an equilibrium constant, based on fugacities and assuming an ideal gas mixture, different from the true constant K_f. By definition $f_i' = \gamma_i' \cdot P$, where P is the total pressure of the mixture and γ_i' is the fugacity coefficient for component i at pressure P.

Here K_f' may be written as

$$K_f' = \frac{x_C^c x_D^d}{x_A^a x_B^b} \cdot \frac{\gamma_C'^c \gamma_D'^d}{\gamma_A'^a \gamma_B'^b} \cdot P^{\Delta n}, \tag{1.54}$$

where $\Delta n = (c+d)-(a+b)$.

Equation (1.54) is readily applied to the calculation of equilibrium mole fractions, since values of γ' can be obtained from generalised fugacity coefficient charts which display γ' as a function of the reduced temperature (T/T_c) and reduced pressure (p/p_c) (see Fig. 1.2).

In the case where all gases are assumed to be ideal, $\gamma_i' = 1$ and (1.54) is written as

$$K_p' = \frac{x_C^c x_D^d}{x_A^a x_B^b} \cdot P^{\Delta n}. \tag{1.55}$$

The value used for K_f' is that calculated from the standard free energy change for the reaction at the required temperature.

An idea of the likely range of pressures over which the assumptions of (a) ideal gases and (b) an ideal gas mixture hold, can be obtained by examining the constancy of calculated values of K_p' and K_f' for the synthesis of NH_3 at 723°K. (Table 1.1).

TABLE 1.1

P atm	$K_p' \times 10^3$	K_γ'	$K_f' \times 10^3$
10	6·59	0·988	6·51
30	6·76	0·969	6·55
50	6·90	0·953	6·57
100	7·25	0·905	6·57
300	8·84	0·750	6·63
600	12·94	0·573	7·42
1000	23·23	0·443	10·32

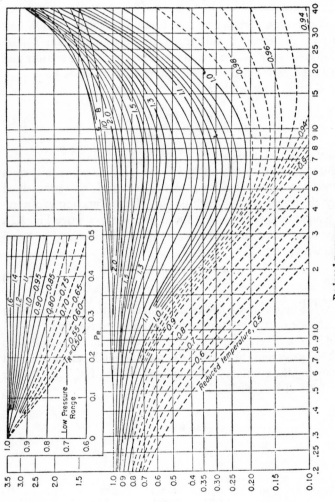

Reduced pressure

Fig. 1.2. Fugacity coefficient of pure compounds as a function of reduced temperature and pressure

It may be concluded that the ideal gas assumption is really valid up to only about 50 atm whereas the gas mixture would appear to be ideal to approximately 300 atm.

To facilitate the calculation of moles of products in equilibrium with reactants, equation (1.54) may be written in the following form

$$\frac{n_C{}^c n_D{}^d}{n_A{}^a n_B{}^b} = \frac{K_f{}'}{K_\gamma{}'} \cdot \left(\frac{n_C + n_D + n_A + n_B + n_I}{P}\right)^{\Delta n} \qquad (1.56)$$

where n represents the number of moles of each reacting species, and n_I has been introduced to take account of the moles of inert gases which might be present.

1.6.1.1 *Effect of Reaction Conditions on Equilibrium Conversions*

The effect of variation of different parameters on the equilibrium conversions can be deduced by examining equation (1.56).

(a) *Effect of temperature variation.* Increase in $K_f{}'$ leads to an increase in product yields. Hence for an endothermic reaction equilibrium conversion increases with rise in temperature, but decreases for an exothermic reaction. Thus for exothermic reactions the theoretical yields are highest at low temperatures, where unfortunately the rates of formation of products will be less than at higher temperatures.

(b) *Effect of pressure variation.* The true equilibrium constant, K_f, is not affected by pressure when using the standard state condition of unit fugacity. However $K_f{}'$ does vary with pressure to some extent as is seen in Table 1.1, because the assumption regarding an ideal gas mixture is not valid over the whole pressure range. In equation (1.56) the value of $K_f{}'$, is taken as constant, since it is calculated from $\Delta G°$, which is independent of pressure, but $K_\gamma{}'$ varies with pressure.

The equilibrium conversion is however, markedly dependent upon pressure when the reaction produces a change in the number of moles during reaction. In the case when Δn is negative, it can be seen from (1.56) that an increase in total pressure leads to an increase in moles of products formed, but the opposite is the case for an increase in pressure when Δn is positive. The rule of thumb suggestion that if $\Delta G°$ is negative, then the reaction is promising, may be seen to depend to a great extent on the overall pressure. In particular, for a reaction at high pressure in which Δn is negative and which has a value of $K_f{}'$ less than 1, corresponding to $\Delta G° > 0$, a satisfactory

B

product yield may still be obtained if the value of P is high enough to offset the low K_f'. A case in point is the high temperature, high pressure industrial synthesis of methanol, where at the temperature and pressure of synthesis (658°K and 340 atm) the value of $\Delta G°$ is about 44000 J/mol.

(c) *Effect of dilution with inert gas.* The addition of inerts to the system requires that an extra term, n_I, be added to the numerator on the right-hand side of (1.56). The effect of an increase in n_I is similar to that brought about by a decrease in total pressure. If the reaction produces an increase in the number of moles of gaseous components, the equilibrium conversion is increased by dilution with inert gas. If $\Delta n = 0$, addition of inert gas has no effect on conversion.

(d) *Effect of excess reactants.* By using an excess of one reactant it is possible to drive the equilibrium to the product side, but with the disadvantage that the product must be recovered from a mixture which contains one reactant in large excess. In fact the highest concentration of product will be found when stoichiometric amounts of reactants are used.

(e) *Effect of presence of products in initial reacting system.* Obviously if products are already present in the reacting system the effective conversion to product will be reduced, because the equilibrium conversion is constant irrespective of the composition of feed to the reactor.

1.6.1.2 *Computation of Equilibrium Compositions*

In most cases other than those involving the simplest of stoichiometry the equilibrium equation will be complex, involving solution to a cubic, quartic or higher order equation. An example is in the synthesis of ammonia from hydrogen and nitrogen where the stoichiometric equation is written as

$$N_2 + 3H_2 \rightleftharpoons 2NH_3.$$

The equilibrium equation, as given by (1.54) is

$$K_f' = \frac{(n_{NH_3})^2}{(n_{N_2})(n_{H_2})^3} \cdot K_\gamma' \left(\frac{P}{n_{N_2} + n_{H_2} + n_{NH_3}}\right)^{-2}.$$

Now if we start with the stoichiometric mixture of reactants, at equilibrium there will be $1 - X$ moles of N_2, $3(1 - X)$ of H_2 and $2X$ of ammonia, where X is the degree of conversion. The total number of moles is then $4-2X$. Hence

$$K_f' = \frac{(2X)^2}{(1-X)(3-3X)^3} \cdot \left(\frac{4-2X}{P}\right)^2$$

where K_γ' is taken as unity. To evaluate X it can be seen that a quartic equation must be solved.

Solving such equations is time-consuming unless performed by computer methods. In this respect mention should be made of the generalised computer solutions contained in the handbook by Noddings and Mullet (1965). The book provides tabulated solutions for conversion and equilibrium composition for an estimated 90% of possible reactions involving stoichiometric reactants of mole number up to five. These tables may be applied to equilibria involving real as well as ideal gases, at any value of total pressure and both with and without addition of inert gases. Solutions may also be obtained for reactant mixtures containing other than the stoichiometric amounts.

Further details of the use of these tables are to be found in the introduction to the handbook.

Example 1.4. Examine the effect of variation of temperature, pressure and reactant ratio on conversion in the vapour phase hydration of ethylene, by calculating the ethanol yield under the following conditions of reaction

	Temp., °K	Pressure atm.	Water/ethylene mole ratio
(a)	523	50	1/1
(b)	593	50	1/1
(c)	593	100	1/1
(d)	593	50	5/1

$$\log K_f' = \frac{2132}{T} - 6 \cdot 241$$

Critical Data

	T_c, °K	p_c, atm
C_2H_4	282·8	50·9
H_2O	647·3	218·5
C_2H_5OH	516·2	63·1

Solution. From the data given,

at 523°K, $\log K_f' = \dfrac{2132}{523} - 6.241 = -2.164$

and $K_f' = 6.855 \times 10^{-3}$

and at 593°K, $\log K_f' = \dfrac{2132}{593} - 6.241 = -2.646$

and $K_f' = 2.259 \times 10^{-3}$.

Equation (1.56) can be used for evaluation of conversion

$$\frac{n_{C_2H_5OH}}{(n_{C_2H_4})(n_{H_2O})} = \frac{K_f'}{K_\gamma'}\left(\frac{n_{C_2H_5OH}+n_{C_2H_4}+n_{H_2O}}{P}\right)^{\Delta n}.$$

(a) and (b)

Let X moles ethanol be formed/mole of C_2H_4 initially present.

$$C_2H_4 + H_2O \rightleftharpoons C_2H_5OH$$

At equilibrium

No. of moles $1-X$ $1-X$ X

K_γ' must be calculated from fugacity coefficients for each species.

	523°K, 50 atm			593°K, 50 atm		
	T_R	p_R	γ'	T_R	p_R	γ'
C_2H_4	$\dfrac{523}{282.8} = 1.85$	$\dfrac{50}{50.9} = 1$	0.98	$\dfrac{593}{282.8} = 2.1$	1	0.985
C_2H_5OH	$\dfrac{523}{516.2} = 1.01$	$\dfrac{50}{63.1} = 0.79$	0.66	$\dfrac{593}{516.2} = 1.15$	0.79	0.81
H_2O	$\dfrac{523}{647.3} = 0.81$	$\dfrac{50}{218.5} = 0.23$	0.82	$\dfrac{593}{647.3} = 0.92$	0.23	0.88

$$K_\gamma' = \frac{0.66}{(0.98)(0.82)} \qquad\qquad K_\gamma' = \frac{0.81}{(0.985)(0.88)}$$

$$= 0.82 \qquad\qquad\qquad\qquad = 0.935$$

At 523°K, 50 atm, water/ethylene mole ratio = 1/1

$$\frac{X}{(1-X)(1-X)} = \frac{0.00685}{0.82}\left(\frac{2-X}{50}\right)^{-1}$$

Conversion = 16.0%.

At 593°K, 50 atm, mole ratio = 1/1

$$\frac{X}{(1-X)(1-X)} = \frac{0.00226}{0.935}\left(\frac{2-X}{50}\right)^{-1}$$

Conversion = 5.6%.

(c)

	593°K, 100 atm		
	T_R	p_R	γ'
C_2H_4	2.1	2	0.97
C_2H_5OH	1.15	1.58	0.575
H_2O	0.92	0.46	0.74

$$K_\gamma' = \frac{0.575}{(0.97)(0.74)} = 0.80$$

At 593°K, 100 atm, mole ratio = 1/1

$$\frac{X}{(1-X)(5-X)} = \frac{0.00226}{0.80}\left(\frac{2-X}{100}\right)^{-1}$$

Conversion = 11.6%.

(d) At $T = 593$°K, 50 atm, water/ethylene mole ratio = 5/1

$$C_2H_4 + H_2O \rightleftharpoons C_2H_5OH$$

No. of moles $1-X$ $5-X$ X

$$\frac{X}{(1-X)(5-X)} = \frac{0.00226}{0.935}\left(\frac{6-X}{50}\right)^{-2}$$

Conversion = 9.1%

1.6.2   HOMOGENEOUS SYSTEM—MULTIPLE GAS PHASE REACTIONS

In the previous section the procedure for the evaluation of equilibrium compositions when only a single reaction is involved was developed. However in many instances several equations must be postulated in order to account for all the products present at equilibrium. For such a system the equilibrium pertaining to each reaction must be taken into account.

Suppose that the following three gas reactions are involved and that equilibrium is established:

$$a\text{A} + b\text{B} \rightleftharpoons c\text{C} + d\text{D} \tag{1.57}$$

$$a\text{A} + c\text{C} \rightleftharpoons e\text{E} \tag{1.58}$$

$$b\text{B} + f\text{F} \rightleftharpoons g\text{G} + h\text{H}. \tag{1.59}$$

Assuming that the gases are ideal, the equilibrium constants for the three reactions may be written

$$K_1 = \frac{x_C^c x_D^d}{x_A^a x_B^b} P^{(c+d-a-b)} \tag{1.60}$$

$$K_2 = \frac{x_E^e}{x_A^a x_C^c} P^{(e-a-b)} \tag{1.61}$$

$$K_3 = \frac{x_G^g x_H^h}{x_B^b x_F^f} P^{(g+h-b-f)}. \tag{1.62}$$

A fourth equation could be deduced from these three by addition

$$2a\text{A} + 2b\text{B} + f\text{F} \rightleftharpoons d\text{D} + e\text{E} + g\text{G} + h\text{H} \tag{1.63}$$

and the equilibrium constant for this reaction would be given by

$$K_4 = \frac{x_D^d x_E^e x_G^g x_H^h}{x_A^{2a} x_B^{2b} x_F^f} P^{(d+e+g+h-2a-2b-f)}. \tag{1.64}$$

However this equation will not be independent of the other three and, in addition, if it alone were used to calculate the composition of the equilibrium mixture, no result would be obtained for component C. In order to evaluate the equilibrium amount of each component the three equations, (1.60)–(1.62), must be solved simultaneously. The general rule which applies is that the number of independent reactions which must be considered is equal to the least number of equations which includes every reactant and product which is present to an appreciable extent in the equilibrium mixture.

Example 1.5. Calculate the equilibrium gas composition obtained when methane and steam, in the mole ratio of 1/2, are passed over a catalyst at 773°K and 1 atm pressure. It may be assumed that the following two reactions account for all the species present at equilibrium

$$CH_4 + 2H_2O \rightleftharpoons CO_2 + 4H_2 \qquad (a)$$

$$CO_2 + H_2 \rightleftharpoons CO + H_2O. \qquad (b)$$

At 773°K, $K_a = 0.046$ and $K_b = 0.205$.

Solution. Writing the extent of conversion in (a) as X and that in (b) as Y, the number of moles of each component present at equilibrium is given as follows

$$CH_4 + 2H_2O \rightleftharpoons CO_2 + 4H_2$$

No. of moles $1 - X \quad 2 - 2X + Y \quad X - Y \quad 4X - Y$

$$CO_2 + H_2 \rightleftharpoons CO + H_2O$$

No. of moles $X - Y \quad 4X - Y \quad Y \quad 2 - 2X + Y$

$$n_{CH_4} = 1 - X, \quad n_{H_2O} = 2 - 2X + Y, \quad n_{CO_2} = X - Y,$$

$$n_{H_2} = 4X - Y, \quad n_{CO} = Y$$

Total number of moles, $\Sigma n = 3 + 2X$.

It should be noted that these reactions cannot be considered in isolation from one another and of course the equilibrium concentration of a component is the same in each reaction.

Two equations only are required to enable a complete solution for the equilibrium composition to be made according to the general rule stated above.

$$K_a = 0.046 = \frac{(X - Y)(4X - Y)^4}{(1 - X)(2 - 2X + Y)^2} \left(\frac{1}{3 + 2X}\right)^2$$

$$K_b = 0.205 = \frac{(Y)(2 - 2X + Y)}{(X - Y)(4X - Y)}$$

Solving these two simultaneous equations gives

$$X = 0.336 \qquad Y = 0.054$$

(Discussion of the technique of solution of this type of problem will be outlined in Chapter 4).

The number of moles of each component at equilibrium will therefore be

$$n_{CH_4} = 1 - X \qquad = 0 \cdot 664$$

$$n_{H_2O} = 2 - 2X + Y = 1 \cdot 382$$

$$n_{CO_2} = X - Y \qquad = 0 \cdot 282$$

$$n_{H_2} = 4X - Y \qquad = 1 \cdot 290$$

$$n_{CO} = Y \qquad = 0 \cdot 054$$

$$\text{Total moles, } \Sigma n = 3 + 2X \qquad = 3 \cdot 672$$

The percentage composition is

$$CH_4 = 18 \cdot 08 \quad H_2O = 37 \cdot 64 \quad CO_2 = 7 \cdot 68 \quad H_2 = 35 \cdot 13$$
$$CO = 1 \cdot 47$$

1.6.3 Homogeneous System—Reactions in Liquid Phase

For the reaction

$$aA + bB \rightleftharpoons cC + dD$$

equation (1.34), in terms of activities, gives a perfectly general representation of the equilibrium. For any type of reaction, the form of the activity expression, which depends on the choice of standard state, is crucial to the solution of the equilibrium expression.

For a liquid mixture the standard state often chosen is that of the pure liquid at the reaction temperature and 1 atm. pressure. If the solution is an ideal one

$$f_i = x_i f_i' \qquad (1.65)$$

where f_i is the fugacity of component i in the solution having mole fraction x_i, and f_i' is the fugacity of the pure component at the reaction pressure. But, since pressure has little effect on activities of liquids, we can write

$$f_i = x_i f_i^\circ \qquad (1.66)$$

where f_i° is the pure component fugacity at 1 atm. i.e. the standard state fugacity. For this situation then the activity will be given by

$$a_i = \frac{f_i}{f_i^\circ} = \frac{x_i f_i^\circ}{f_i^\circ} = x_i \qquad (1.67)$$

For an ideal liquid mixture equation (1.34) can therefore be written as

$$K = \frac{(x_C)^c (x_D)^d}{(x_A)^a (x_B)^b}. \qquad (1.68)$$

For a non-ideal liquid mixture we can write

$$f_i = \gamma_i x_i f_i^\circ \qquad (1.69)$$

where γ_i is a fugacity coefficient. In this case therefore

$$a_i = \gamma_i x_i. \qquad (1.70)$$

This expression for activity can very rarely be applied because of the difficulties involved in evaluating γ_i.

It must be remembered that the value of K has to be calculated from ΔG° data relating to the appropriate standard state being employed.

1.6.4 HETEROGENEOUS REACTIONS

A type of reaction which can be simply treated is that involving solids or liquids, present in the pure state, as well as gases, i.e. where there is no solution formation. A typical example would be that of a solid material decomposing into solid and gaseous products as in the case of calcium carbonate, which forms calcium oxide and carbon dioxide on decomposition:

$$CaCO_3(s) \rightleftharpoons CaO(s) + CO_2(g).$$

For this example

$$K = \frac{a_{CaO} a_{CO_2}}{a_{CaCO_3}}. \qquad (1.72)$$

The standard states chosen for the solids or liquids in this system are those of the pure materials at the reaction temperature and 1 atm pressure, i.e. the activities of the solids and liquids are unity under these conditions.

The pressure of the system may very likely differ from 1 atm. in which case the activities will deviate from unity; however, activities of pure liquids and solids are not very pressure dependent, and hence equation (1.72) can be simplified to

$$K = a_{CO_2} = f_{CO_2} = \gamma p_{CO_2} \tag{1.73}$$

if the standard state of unit activity for the gas is that of pure gas at the temperature and pressure of the reaction system.

REFERENCES

AMERICAN PETROLEUM INSTITUTE. 1952. Selected values of properties of hydrocarbons and related compounds. *Amer. Petrol. Inst. Res. Proj.* 44, Carnegie Inst. Tech., Pittsburgh.

DODGE, B. F. 1944. *Chemical engineering thermodynamics.* McGraw-Hill, New York.

HOUGEN, O. A., WATSON, K. M., and RAGATZ, R. A. 1959. *Chemical process principles: part 2,* 2nd Ed. Wiley, New York.

JANAF. 1965. Thermochemical tables. *Pubn PB* 168–370. Clearing House for Federal Scientific and Technical Information, Springfield, Virginia.

NODDINGS, C. R., and MULLET, G. M. 1965. *Handbook of compositions at thermodynamic equilibrium.* Wiley Interscience, New York.

SMITH, J. M., and VAN NESS, H. C. 1959, 2nd Ed. *Introduction to chemical engineering thermodynamics.* McGraw-Hill, New York.

WAGMAN, D. D. (ed). 1952. *Selected values of chemical thermodynamic properties.* Natl. Bureau of Stds., Circ. 500.

2
KINETICS AND REACTOR DESIGN

2.1 Introduction

Equilibrium thermodynamics is used to predict the extent to which a chemical reaction (or reactions) proceeds, but it is the subject of kinetics which must be applied to determine the rates at which reactions take place. The rate of reaction determines the size of reactor required to process a given amount of reactant per unit time, and is therefore a vitally important factor to be considered in the design of a reactor. The intention of this chapter is solely to outline in a general fashion some concepts of kinetics which are relevant to the subject matter of the following chapters, where the necessary kinetic aspects will be considered with the minimum of background detail. Readers are strongly recommended to consult one of the excellent text books on kinetics and reactor design, such as those by Cooper and Jeffreys (1971), Smith (1970) and Levenspiel (1962), for a fuller appreciation of the subject matter considered here.

2.2 Rate Equations for Reactions in Batch Systems

The differential rate expression for a reaction in a constant volume batch system is of the form

$$(-r_A) = -(dC_A/dt) = kf(C_A{}^{\alpha}C_B{}^{\beta}\ldots), \qquad (2.1)$$

where $(-r_A) = -(dC_A/dt)$ is the rate of reaction expressed as the rate of disappearance of reactant A, k is the rate constant, the value of which depends on temperature, while $f(C_A{}^{\alpha}C_B{}^{\beta}\ldots)$ represents the function of concentrations of reactants, and sometimes of products, on which the rate depends. α and β are termed the order of reaction for A and B respectively.

Differential rate expressions are normally obtained by experimental determination of reactant or product concentration change as a function of time, temperature and nature and condition of catalyst, if used. Rate expressions are very often empirical in nature, but occasionally they may be determined theoretically giving rise, in certain instances, to rather complex expressions involving more than one rate constant, if the reaction proceeds by a number of steps in which reactive intermediates take part. For design use rate expressions which are as simple as possible are desirable, so that empirical forms are very often used.

2.2.1 IRREVERSIBLE FIRST ORDER REACTION

For the first order reaction, A → products, taking place in a non-flow, constant volume reactor, the rate expression may be written

$$-(dC_A/dt) = kC_A \qquad (2.2)$$

where the order, α, equals unity.

Now $C_A = C_{A_0}(1 - X_A)$ where C_A is the concentration at time t, C_{A_0} is initial concentration and X_A is conversion. Differentiation gives $-dC_A/dt = C_{A_0}(dX_A/dt)$ and hence, in terms of conversion, equation (2.2) becomes

$$C_{A_0}(dX_A/dt) = kC_{A_0}(1 - X_A)$$

or

$$(dX_A/dt) = k(1 - X_A). \qquad (2.3)$$

Equation (2.2) is integrated to give

$$C_A = C_{A_0}e^{-kt} \quad \text{or} \quad \ln(C_A/C_{A_0}) = -kt. \qquad (2.4)$$

The integrated form of equation (2.3) is

$$-\ln(1 - X_A) = kt. \qquad (2.5)$$

From equation (2.4) it may be seen that the rate constant will have units of $(\text{time})^{-1}$.

For reaction under non-constant volume conditions, e.g. under constant pressure conditions, the rate is written

$$(-r_A) = -\frac{1}{V}\frac{dn_A}{dt} = -\frac{1}{V}\frac{d(C_A V)}{dt} = -\frac{1}{V}\left(\frac{V dC_A}{dt} + \frac{C_A dV}{dt}\right)$$

$$= -\frac{dC_A}{dt} - \frac{C_A}{V}\cdot\frac{dV}{dt}, \qquad (2.6)$$

i.e. a second term must be added to the constant volume expression. To avoid using this rather cumbersome expression, the assumption is made that the volume of the reacting system is proportional to conversion or $V = V_0(1 + E_A X_A)$, where E_A is the fractional change in volume of the system between no conversion and complete conversion. Thus E_A is defined as

$$E_A = \frac{V_{X_A = 1} - V_{X_A = 0}}{V_{X_A = 0}}$$

Now since $n_A = n_{A_0}(1 - X_A)$ and $C_A = (n_A/V)$, C_A is given by the expression

$$C_A = \frac{C_{A_0}(1 - X_A)}{(1 + E_A X_A)}. \tag{2.7}$$

The reaction rate $-dC_A/dt$ then becomes

$$\frac{C_{A_0}}{1 + E_A X_A} \cdot \frac{dX_A}{dt}.$$

For a first order reaction, equation (2.2) is written as

$$\frac{C_{A_0}}{1 + E_A X_A} \cdot \frac{dX_A}{dt} = \frac{kC_{A_0}(1 - X_A)}{1 + E_A X_A}$$

or

$$\frac{dX_A}{dt} = k(1 - X_A), \tag{2.3}$$

which is the same expression as found for a first order reaction in a constant volume system. However the concentrations would not be the same, for in this case

$$C_A = \frac{C_{A_0}(1 - X_A)}{1 + E_A X_A}$$

but equal to $C_{A_0}(1 - X_A)$ for the constant volume case.

2.2.2 IRREVERSIBLE SECOND ORDER REACTION

For reaction $A + B \rightarrow$ products there are two possible differential rate expressions for the constant volume case

(a)

$$\frac{-dC_A}{dt} = kC_A^2 \tag{2.8}$$

where $C_A = C_B$

and (b)

$$\frac{-dC_A}{dt} = kC_A C_B \tag{2.9}$$

where the reaction is first order with respect to each reactant.

In terms of conversion these are written as

$$\frac{dX_A}{dt} = kC_{A_0}(1 - X_A)^2 \tag{2.10}$$

and

$$\frac{dX_A}{dt} = kC_{A_0}(1 - X_A)\left(\frac{C_{B_0}}{C_{A_0}} - X_A\right) \tag{2.11}$$

since $C_{A_0}X_A = C_{B_0}X_B$.

Equation (2.8) is integrated to give

$$kt = \left(\frac{1}{C_A} - \frac{1}{C_{A_0}}\right) \tag{2.12}$$

where C_{A_0} is concentration of A at time $t = 0$.

Similarly (2.10) is integrated to give

$$\frac{1}{C_{A_0}} \cdot \frac{X_A}{1 - X_A} = kt. \tag{2.13}$$

The integral expression for (2.11) is

$$\int_0^{X_A} \frac{dX_A}{(1 - X_A)(Z - X_A)} = C_{A_0}k \int_0^t dt \quad \text{where} \quad Z = C_{B_0}/C_{A_0}.$$

Now splitting into partial fractions the result is

$$\int_0^{X_A} \frac{dX_A}{(Z - 1)(1 - X_A)} + \int_0^{X_A} \frac{dX_A}{(1 - Z)(Z - X_A)} = C_{A_0} kt,$$

which on integration gives

$$-\ln(1 - X_A) + \frac{\ln(Z - X_A)}{Z} = (Z - 1)C_{A_0} kt$$

or

$$\ln \frac{Z - X_A}{Z(1 - X_A)} = (Z - 1)C_{A_0} kt$$

or

$$\ln \frac{(C_{B_0} - C_{A_0} X_A)}{C_{B_0}(1 - X_A)} = \ln \frac{(1 - X_B)}{(1 - X_A)} = (C_{B_0} - C_{A_0})kt$$

or

$$\ln \frac{C_B C_{A_0}}{C_{B_0} C_A} = (C_{B_0} - C_{A_0})kt.$$

$$(2.14)$$

For a variable volume reaction the form of equation (2.10) is

$$\frac{C_{A_0}}{1 + E_A X_A} \cdot \frac{dX_A}{dt} = \frac{kC_{A_0}^2(1 - X_A)^2}{(1 + E_A X_A)^2}.$$

or

$$\frac{dX_A}{dt} = \frac{kC_{A_0}(1 - X_A)^2}{(1 + E_A X_A)}.$$

$$(2.15)$$

On integrating by the method of partial fractions we have

$$\int_0^{X_A} \frac{(1 + E_A X_A)dX_A}{(1 - X_A)^2} = \int_0^{X_A} \frac{(1 + E_A)dX_A}{(1 - X_A)^2} - \int_0^{X_A} \frac{E_A dX_A}{1 - X_A} = kC_{A_0}t$$

or

$$\frac{(1 + E_A)X_A}{(1 - X_A)} + E_A \ln(1 - X_A) = kC_{A_0}t.$$

$$(2.16)$$

2.2.3 REVERSIBLE REACTIONS

Consider the general reaction $A + B \rightleftharpoons C + D$.

If the reaction is first order with respect to each species, the rate equation which applies is

$$\frac{-dC_A}{dt} = \frac{dC_C}{dt} = k_1 C_A C_B - k_2 C_C C_D$$

$$(2.17)$$

where k_1 is the rate constant for the forward reaction and k_2 for the reverse. At equilibrium

$$\frac{-dC_A}{dt} = \frac{dC_C}{dt} = 0$$

and

$$\frac{k_1}{k_2} = \frac{C_C C_D}{C_A C_B} = K \text{ (equilibrium constant)}.$$

First Order Reversible Reaction. The rate expression for reaction $A \rightleftharpoons B$ is

$$\frac{-dC_A}{dt} = \frac{dC_B}{dt} = k_1 C_A - k_2 C_B$$

$$= k_1 C_{A_0}(1 - X_A) - k_2 C_{A_0} X_A \qquad (2.18)$$

if the concentration of B is zero initially.

Now at equilibrium

$$\frac{k_1}{k_2} = \frac{C_{Be}}{C_{Ae}} = \frac{X_{Ae}}{(1 - X_{Ae})}$$

where subscript e represents the equilibrium condition. Equation (2.18) then becomes

$$\frac{-dC_A}{dt} = k_1 C_{A_0}(1 - X_A) - \frac{k_1 C_{A_0}(1 - X_{Ae})}{X_{Ae}} \cdot X_A$$

or

$$\frac{-dC_A}{dt} = k_1 C_{A_0} \left\{ \frac{X_{Ae} - X_A X_{Ae} - X_A + X_A X_{Ae}}{X_{Ae}} \right\}$$

$$= k_1 C_{A_0}[1 - (X_A/X_{Ae})]. \qquad (2.19)$$

Alternatively, since

$$\frac{-dC_A}{dt} = C_{A_0} \frac{dX_A}{dt},$$

equation (2.19) can be written as

$$\frac{dX_A}{dt} = k_1 \left(1 - \frac{X_A}{X_{Ae}} \right). \qquad (2.20)$$

Integration of equation (2.20) gives

$$\int_0^{X_A} \frac{dX_A}{[1 - (X_A/X_{Ae})]} = k_1 t$$

or

$$-X_{Ae} \ln [1 - (X_A/X_{Ae})] = k_1 t \qquad (2.21)$$

or, by rearrangement,

$$X_A = X_{Ae}[1 - \exp(-k_1 t/X_{Ae})]. \qquad (2.22)$$

Alternatively, since

$$X_{Ae} = k_1/(k_1+k_2),$$

$$\ln\left(\frac{X_{Ae}}{X_{Ae}-X_A}\right) = (k_1+k_2)t. \qquad (2.23)$$

2.2.4 CONSECUTIVE REACTIONS

Many instances exist of reactions involving the simultaneous formation and decomposition of a required species. For example, in many oxidation processes the oxygenated product has a tendency to degrade further to carbon dioxide and water.

A general scheme may be written for reactions of this type

$$A \xrightarrow{k_1} B \xrightarrow{k_2} C \text{ etc.}$$

where B is the required intermediate in a two-stage reaction.

The simplest case to consider is that where both k_1 and k_2 are first order rate constants. The rate equations for the various species in a constant volume system are

$$dC_A/dt = -k_1C_A \qquad (2.2)$$

$$dC_B/dt = k_1C_A-k_2C_B \qquad (2.24)$$

$$dC_C/dt = k_2C_B \qquad (2.25)$$

where $C_{B_0} = C_{C_0} = 0$.

The integral form of equation (2.2) is equation (2.4)

$$C_A = C_{A_0} \exp(-k_1t). \qquad (2.4)$$

Substitution in equation (2.24) gives the first order linear differential equation

$$(dC_B/dt)+k_2C_B = k_1C_{A_0} \exp(-k_1t). \qquad (2.26)$$

This equation is of the general form $(dy/dx)+Py = Q$ which has the solution

$$y\exp(\int Pdx) = \int Q \exp(\int Pdx)\,dx+\text{constant}.$$

Hence the solution to equation (2.26) is

$$C_B \exp(k_2t) = k_1C_{A_0} \int \exp(-k_1t) \exp(k_2t)\,dt+\text{constant}$$

$$= k_1C_{A_0} \frac{\exp(k_2-k_1)t}{k_2-k_1}+\text{constant}. \qquad (2.27)$$

To evaluate the constant, apply the boundary condition that at $t = 0$, $C_B = 0$. We then obtain

$$\text{constant} = \frac{-k_1}{k_2 - k_1} C_{A_0}.$$

Substitution in equation (2.27) and rearrangement gives

$$\frac{C_B}{C_{A_0}} = \frac{k_1}{k_2 - k_1} [\exp(-k_1 t) - \exp(-k_2 t)]. \qquad (2.28)$$

C_C may be obtained by a mass balance.

For a constant volume system

$$C_{A_0} = C_A + C_B + C_C.$$

Hence, using equations (2.4) and (2.28) to substitute for C_A and C_B respectively, the following expression is obtained for C_C

$$\frac{C_C}{C_{A_0}} = 1 - \exp(-k_1 t) - \frac{k_1}{k_2 - k_1} \{\exp(-k_1 t) - \exp(-k_2 t)\}$$

$$= 1 + \frac{k_2}{k_1 - k_2} \exp(-k_1 t) + \frac{k_1}{k_2 - k_1} \exp(-k_2 t). \qquad (2.29)$$

The variation of concentration with time for each species is shown in Fig. 2.1. It can be seen that C_B/C_{A_0} passes through a maximum. The position of this maximum point can be determined by differentiating C_B with respect to time and setting the differential equal to zero. Thus, differentiating equation (2.28) and equating to zero gives

$$dC_B/dt = -k_1 \exp(-k_1 t_{max}) + k_2 \exp(-k_2 t_{max}) = 0.$$

From this expression t_{max}, the time at which the concentration of B reaches a maximum, is then given by

$$t_{max} = \frac{\ln(k_2/k_1)}{k_2 - k_1}. \qquad (2.30)$$

Substituting for $\exp(-k_2 t_{max})$ in terms of $\exp(-k_1 t_{max})$ in equation (2.28) gives

$$\frac{C_{Bmax}}{C_{A_0}} = \frac{k_1}{k_2 - k_1} \left(\exp(-k_1 t_{max}) - \frac{k_1}{k_2} \exp(-k_1 t_{max}) \right)$$

$$= \frac{k_1}{k_2} \exp(-k_1 t_{max}),$$

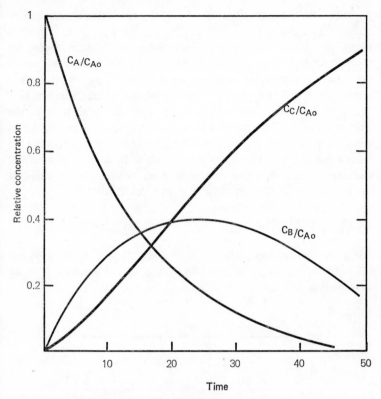

FIG. 2.1. Concentration changes in consecutive first-order
reactions

while substitution for t_{max} using equation (2.30) gives the final result

$$\frac{C_{Bmax}}{C_{A_0}} = \left(\frac{k_1}{k_2}\right) \exp\left[\{-k_1/(k_2-k_1)\} \ln (k_2/k_1)\right]$$

$$= \left(\frac{k_1}{k_2}\right)^{k_2/(k_2-k_1)}. \tag{2.31}$$

2.3 Design Equations for Flow Reactors

2.3.1 DESIGN EQUATIONS FOR FLOW REACTORS—ISOTHERMAL

In industrial processing it is very common for reactions to be carried
out under flow conditions. In order to calculate the size of reactor

required, the rate equations derived in the previous sections can be inserted into the appropriate design equation for the flow reactor being considered. The two most common types of flow reactor are those described as plug flow or tubular, and the continuous stirred tank. In the former the reactants are pictured as passing down the reactor tube, which is generally long in relation to its diameter, as if in single file. Each reactant spends exactly the same time in the reactor as any other, the velocity profile is flat and no mixing is assumed to take place. In a stirred tank, on the other hand, complete mixing is assumed and there is a spectrum of residence times for the reacting species. A mean residence time, defined as the ratio of reactor volume/volumetric flow rate, can be calculated.

2.3.1.1 *Plug Flow Reactor*

For a reaction taking place under isothermal conditions, a design equation may be deduced by considering a mass balance over an increment of volume, dV (Fig. 2.2).

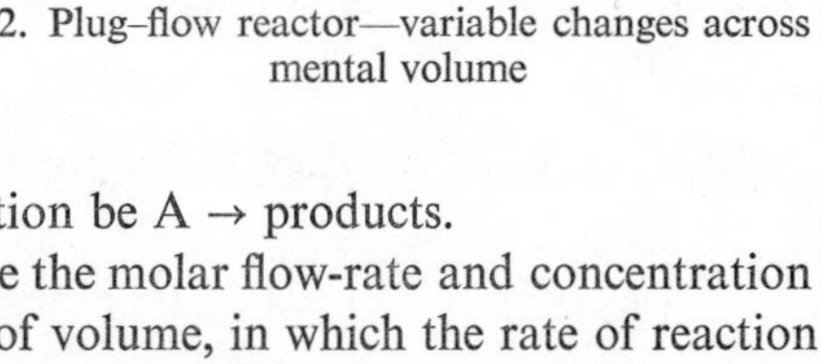

$$
\begin{array}{c|c|c}
 & dV & \\
\hline
F_A & & F_A + dF_A \\
C_A \;\rightarrow & (-r_A) \;\rightarrow & C_A + dC_A \\
X_A & & X_A + dX_A \\
\end{array}
$$

Fig. 2.2. Plug–flow reactor—variable changes across incremental volume

Let the reaction be A → products.

Let F_A, C_A be the molar flow-rate and concentration of A entering the increment of volume, in which the rate of reaction is $(-r_A)$.

Let $F_A + dF_A$, $C_A + dC_A$ be the molar flow-rate and concentration of A leaving the increment.

Hence a mass balance for reactant gives

$$F_A = F_A + dF_A + (-r_A) \cdot dV \qquad (2.32)$$

from which

$$-dF_A = (-r_A)dV. \qquad (2.33)$$

Now

$$F_A = F_{A_0}(1 - X_A) \quad \text{and} \quad dF_A = -F_{A_0} \cdot dX_A.$$

Substitution for dF_A in equation (2.33) gives

$$F_{A_0} \cdot dX_A = (-r_A) \cdot dV$$

or

$$\frac{dV}{F_{A_0}} = \frac{dX_A}{(-r_A)}. \tag{2.34}$$

The integrated form of equation (2.34) is

$$\frac{V}{F_{A_0}} = \int_0^{X_A} \frac{dX_A}{(-r_A)} \tag{2.35}$$

where X_A is conversion of A at the reactor exit, assuming no conversion initially.

If both sides of this equation are multiplied by C_{A_0} we obtain

$$\tau = \frac{V}{F_{A_0}} \cdot C_{A_0} = C_{A_0} \int_0^{X_A} \frac{dX_A}{(-r_A)}. \tag{2.36}$$

τ is called the space time. Its value will depend upon the temperature and pressure conditions under which C_{A_0} is measured. This is particularly important in the case of gaseous reactions where concentration is a strong function of temperature and total pressure. The space time will not be the same as the residence time in the reactor unless there is no change in number of moles during reaction and the concentration, C_{A_0}, is measured at the isothermal temperature and pressure in the reactor. The reciprocal of the space time is called the space velocity.

Equation (2.36) can be integrated analytically in certain instances by inserting the appropriate rate equation for $(-r_A)$. Thus for a *constant density first order irreversible* reaction the rate equation is

$$(-r_A) = kC_A = kC_{A_0}(1 - X_A) \tag{2.2}$$

which can be substituted in equation (2.36) to give

$$\tau = C_{A_0} \cdot \frac{V}{F_{A_0}} = C_{A_0} \int_0^{X_A} \frac{dX_A}{kC_{A_0}(1 - X_A)}$$

or

$$k\tau = -\ln(1 - X_A). \tag{2.37}$$

It is to be noted that this equation is of the same form as equation (2.5) for the batch reactor. In fact the two equations will be identical if τ is the true residence time in the reactor, i.e. if it has been calculated at the temperature and pressure in the reactor. This means that the conversion in the two types of reactor will be the same if the time spent in each is the same.

For a *first order reversible* reaction

$$(-r_A) = k_1 C_{A_0} [1 - (X_A/X_{Ae})]. \tag{2.19}$$

Substitution for $(-r_A)$ in equation (2.36) gives

$$\tau = C_{A_0} \int_0^{X_A} \frac{dX_A}{k_1 C_{A_0}[1 - (X_A/X_{Ae})]},$$

from which

$$k_1 \tau = -X_{Ae} \ln [1 - (X_A/X_{Ae})]. \tag{2.38}$$

For a *first order irreversible reaction* in which the *density changes*, the design equation becomes (from equation 2.7)

$$\tau = C_{A_0} \cdot \frac{V}{F_{A_0}} = C_{A_0} \int_0^{X_A} \frac{(1 + E_A X_A)dX_A}{k C_{A_0}(1 - X_A)}. \tag{2.39}$$

On splitting into partial fractions and integrating we obtain

$$k\tau = \int_0^{X_A} \frac{(1 + E_A)dX_A}{1 - X_A} - \int_0^{X_A} E_A dX_A$$

or

$$k\tau = (1 + E_A) \ln \frac{1}{1 - X_A} - E_A X_A. \tag{2.40}$$

2.3.1.2   *Stirred Tank Reactor*

The design equation for a stirred tank reactor operating under isothermal conditions is evaluated by carrying out a mass balance over the whole reactor, since in this case the average concentration in the reactor is everywhere the same and equal to that at the reactor outlet. A schematic mass balance is shown in Fig. 2.3.

Flow in = flow out + disappearance by reaction

$$F_{A_0} = F_A + (-r_A)V. \tag{2.41}$$

Since $F_A = v \cdot C_A$ and $F_{A_0} = v \cdot C_{A_0}$, substitution in equation (2.41) gives

$$v \cdot C_{A_0} = v \cdot C_A + (-r_A)V$$

and

$$C_{A_0} = C_A + (-r_A) \cdot \bar{t} \qquad (2.42)$$

where $\bar{t} = V/v$.

Rearrangement of equation (2.42) leads to

$$\bar{t} = \frac{C_{A_0} - C_A}{(-r_A)}. \qquad (2.42)$$

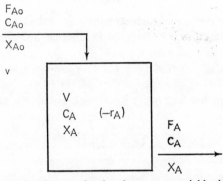

FIG. 2.3. Continuous stirred tank reactor—variable changes across reactor volume

For a constant density *first order irreversible reaction* $(-r_A) = kC_A$. Substituting for $(-r_A)$ in equation (2.42) gives

$$\bar{t} = \frac{C_{A_0} - C_A}{kC_A}$$

or

$$\frac{C_{A_0}}{C_A} - 1 = k\bar{t}$$

or

$$\frac{C_A}{C_{A_0}} = \frac{1}{1 + k\bar{t}}. \qquad (2.43)$$

Eliminating C_A/C_{A_0} by using $C_A = C_{A_0}(1 - X_A)$ results in the expression

$$1 - X_A = \frac{1}{1 + k\bar{t}}$$

or

$$X_A = \frac{k\bar{t}}{1 + k\bar{t}}. \qquad (2.44)$$

For a *first order reversible reaction*

$$(-r_A) = k_1 C_{A_0}[1 - (X_A/X_{Ae})].$$

Substitution in equation (2.42) for $(-r_A)$ in this case and again using the relationship $C_A = C_{A_0}(1 - X_A)$ leads to

$$X_A = \frac{k_1 \bar{t} X_{Ae}}{X_{Ae} + k_1 \bar{t}}. \tag{2.45}$$

The application of the design equations for tubular and continuous stirred tank reactors is considered in Chapter 3 dealing with reversible exothermic reactions.

More generally, stirred tank reactors are not used singly but in series. Thus, the design equation for tank (1), in a series of equal sized reactors in which a first order irreversible reaction is taking place, is

$$\frac{C_{A_1}}{C_{A_0}} = \frac{1}{1 + k\bar{t}}, \tag{2.43}$$

and for tank (2)

$$\frac{C_{A_2}}{C_{A_1}} = \frac{1}{1 + k\bar{t}}.$$

For the *n*th reactor therefore, if the reactors are of the same size,

$$\frac{C_{A_n}}{C_{A_0}} = \frac{C_{A_n}}{C_{A_{n-1}}} \cdots \frac{C_{A_2}}{C_{A_1}} \cdot \frac{C_{A_1}}{C_{A_0}} = \frac{1}{(1 + k\bar{t})^n}. \tag{2.46}$$

Further, rearrangement of equation (2.46), results in

$$1 + k\bar{t} = \left(\frac{C_{A_0}}{C_{A_n}}\right)^{1/n}.$$

Hence the total time spent in *n* reactors is

$$\bar{t}_{tot} = \frac{n}{k} \left\{ \left(\frac{C_{A_0}}{C_{A_n}}\right)^{1/n} - 1 \right\}. \tag{2.47}$$

2.3.2 DESIGN EQUATIONS FOR FLOW REACTORS—NON-ISOTHERMAL

In most industrial reactors the temperature within the reactor is unlikely to be the same as at the entrance, because heat is invariably liberated or consumed during reaction. Hence in calculating the amount of reaction which takes place within a flow reactor, it is necessary that the temperature obtaining at each point within the reactor be properly assessed, since the reaction rate is highly dependent upon temperature. For a tubular reactor the temperature will vary along the length and, in certain circumstances, radially also, but for a stirred tank reactor the temperature within the reactor will be uniform, though its magnitude will depend upon the amount of heat liberated or consumed during reaction.

Two different modes of non-isothermal operation may be considered:

(i) adiabatic, when no heat is exchanged between reactor and surroundings, other than through the transfer of sensible heat to or from reactants and products; and

(ii) where heat is transferred across reactor walls.

Of the two, adiabatic operation is the simpler to treat from a mathematical viewpoint.

2.3.2.1 *Adiabatic Operation of a Tubular Reactor*

There are two different forms of tubular reactor in use (a) a reactor in which homogeneous reactions only take place, i.e. the reactor contents form a single (usually gaseous) phase, and (b) a reactor in which more than one phase is present, usually a solid which acts as a catalyst on which gaseous reaction takes place. Design equations for the two cases are on the whole similar but differences in the formulation will be noted where relevant.

The aim in view in this section is the development of equations which will allow the calculation of species composition and temperature at points along the reactor.

Suppose that the reaction can be written as $A \rightarrow$ products. It is first order with respect to concentration of A, is irreversible and proceeds without change of density. The rate equation may be written

$$(-r_A) = kC_{A_0}(1 - X_A). \tag{2.2}$$

In order to achieve the desired objective it is necessary to postulate a mass and a heat balance over an increment of volume.

Mass Balance.

$$F_{A_0}dX_A = (-r_A)A_c \cdot dZ$$

or

$$\frac{A_c dZ}{F_{A_0}} = \frac{dX_A}{(-r_A)} \tag{2.48}$$

where A_c is the cross sectional area of reactor (m^2) of length Z(m), $(-r_A)$ is the reaction rate in units of kmol/m^3 s, and F_{A_0} is initial flow rate of A in kmol/s. (For a catalytic reaction the mass balance can be written as

$$F_{A_0}dX_A = (-r_{c_A})\rho_B A_c dZ \tag{2.49}$$

where $(-r_{c_A})$ is the reaction rate in units of kmol/kg catalyst s, ρ_B being the catalyst bulk density in units of kg/m^3.)

Energy Balance. Suppose that the reaction is exothermic and the heat liberated is $(-\Delta H)$kJ/kmol of A reacting. The heat liberated will be absorbed by the flowing gas stream. For an increment of length, dZ, the heat balance equation will be

$$(-r_A)A_c dZ(-\Delta H) = \sum (F_i C_{p_i})dT \tag{2.50}$$

where F_i represents the molar flow rate of species i of heat capacity of C_{p_i}(kJ/kmol °K), and dT is the rise in temperature across the increment.

Substitution of equation (2.48) in (2.50) leads to

$$F_{A_0}dX_A(-\Delta H) = \sum (F_i C_{p_i})dT. \tag{2.51}$$

Thus if F_{A_0}, $(-\Delta H)$ and $\sum F_i C_{p_i}$ are known, equation (2.51) may be integrated to give a relationship between X_A and T. For example, for reaction A → B involving A and B only, $\sum (F_i C_{p_i}) = F_{A_0}(1-X_A)C_{pA}+F_{A_0}X_A C_{pB}$ assuming C_p is constant over the temperature range. The integrated form of equation (2.51) is then

$$T = T_1 + \left\{ \frac{(-\Delta H)X_A}{(1-X_A)C_{pA} + X_A C_{pB}} \right\} \tag{2.52}$$

where T_1 is value of T at entrance to reactor, i.e. where $X_A = 0$, and $(-\Delta H)$ is assumed constant over the temperature range considered.

For a given value of X_A and corresponding temperature, T, the rate of reaction $(-r_A)$ can be calculated. The volume of reactor required to achieve a given conversion can be calculated by one of two methods:

(i) by graphical integration of $[1/(-r_A)]$ versus X_A when the area under the curve up to a given conversion, X_A, will equal $A_c Z/F_{A_0}$ from which $V = A_c Z$ can be calculated from a known value of F_{A_0}; or

(ii) by a step-wise numerical procedure where the increment of length, ΔZ, required for a chosen conversion interval, ΔX_A, may be obtained from equation (2.48) applied in a difference form

$$\Delta Z = \frac{F_{A_0}}{(-r_A)_{av} A_c} \Delta X_A \qquad (2.53)$$

where $(-r_A)_{av} = \{$rate at increment entrance (1) + rate at increment exit $(2)\}/2$.

2.3.2.2 Non-Isothermal, Non-Adiabatic Operation of a Tubular Reactor

This type of operation results when heat is either removed or introduced through the reactor walls. In cases where the heat effect within the reactor is not substantial, or when there is little resistance to heat transfer within the reactor or catalyst bed, except at the wall, the temperature at a given axial position may be considered uniform across the radius of the tube. The model describing this sort of behaviour is termed *one-dimensional*, simple differential equations only being invoked in a description. Other assumptions in the simplest form of this model are that the mass transport mechanism involves axial flow only, and that the velocity profile is uniform across the reactor diameter. This form of model is generally applicable in the case of homogeneous reactions and heterogeneous cases involving small heat effects.

In many instances however, where large heat effects are involved, and especially in fixed bed catalytic reactors where heat transfer rates are low because the solid material is generally a poor conductor, large radial temperature gradients may result, and it is necessary to resort to a *two-dimensional* model to describe the temperature and conversion profiles both axially and radially. The equations to be solved are then of a partial differential nature.

Consider the *one-dimensional* case first of all. This model predicts the average temperature and concentration at any cross-section along the reactor and, as in the case of the adiabatic reactor, a mass balance equation, or equations if more than one reaction is involved, and an energy balance are required in addition to the rate equation(s).

Mass balance. Equation (2.48) or (2.49) applies to this case also.

Energy balance. In this case an extra term has to be added to account for the heat transfer across the tube wall. The equation is

$$(-r_{c_A}) \rho_B A_c dZ(-\Delta H) = h_w A_w (T_m - T_w) dZ + \sum F_i C_{p_i} dT \qquad (2.54)$$

where the reaction in this case is assumed to be catalytic. h_w is the wall heat transfer coefficient (kJ/m² s °K), A_w is area of wall/unit length, T_m is mean temperature within bed and T_w is wall temperature (°K).

If $F_{A_0}dX_A$ is substituted for $(-r_{c_A})\rho_B A_c dZ$ the equation becomes

$$F_{A_0}dX_A(-\Delta H) = h_w A_w(T_m - T_w)dZ + \sum F_i C_{p_i}dT. \qquad (2.55)$$

These equations are based on the assumption that there is no resistance to heat transfer except at the wall of the tube. The temperature profile across the reactor cross-section will be as shown in Fig. 2.4.

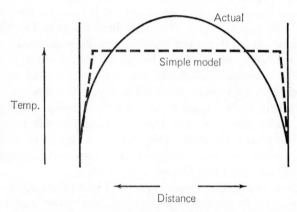

FIG. 2.4. Packed bed reactor—one dimensional model and actual temperature profiles across reactor

An alternative way of writing equation (2.54) is in terms of the overall heat transfer coefficient, U,

$$\frac{dT}{dZ} = \frac{(-r_{c_A})\rho_B A_c(-\Delta H)}{\sum F_i C_{p_i}} - \frac{UA_w}{\sum F_i C_{p_i}}(T_m - T) \qquad (2.56)$$

where T is the temperature of fluid on the outside wall.

Solution for X_A and T as a function of length, Z, is performed by a step-wise trial and error procedure after converting the mass and energy differential equations into difference forms.

A convenient size of conversion increment, ΔX_A, is chosen and a mean temperature, T_m, is assumed. Equation (2.48) is expressed in the following difference form

$$\Delta Z = R_{av}\Delta X_A \qquad (2.53)$$

where R_{av} is a function of the average rate of reaction at the initial temperature and assumed temperature at the end of the increment of conversion ΔX_A. The ΔZ required to obtain the chosen ΔX_A is then calculated.

This value of ΔZ is then substituted into the difference form of equation (2.55)

$$\Delta T = \frac{F_{A0}\Delta X_A(-\Delta H)}{\sum F_i C_{p_i}} - \frac{h_w A_w (T_m - T_w)_{av}}{\sum F_i C_{p_i}} \cdot \Delta Z \qquad (2.57)$$

and the temperature rise, ΔT, computed. This temperature rise should match that calculated on the basis of the assumed temperature i.e. ΔT should equal $\{(T_{m_2} - T_w) - (T_{m_1} - T_w)\}/2$. If not, a new value of T_{m_2} must be chosen and the procedure repeated until agreement is obtained. Obviously this form of calculation is most easily performed by computer. The computation is repeated until the desired value of X_A is obtained.

As stated in the introduction to this section, the two-dimensional approach must be used when radial temperature gradients are likely to be substantial, as in the case of a heterogeneous reaction with a significant heat effect. The temperature and concentrations are now variable in two directions and partial differential equations are necessary for the solution of the model.

Before these equations are derived we shall briefly discuss the mechanism of mass and heat transfer within packed beds. It is almost impossible to describe the hydrodynamics of packed beds because of the irregular arrangement of the packing. The random movement of gas within the packed bed, which is superimposed upon the bulk flow, may be described in terms of an effective diffusivity, D_e, where D_e replaces the true diffusivity in Fick's first law. It has been shown experimentally that the effective diffusivity in the radial and axial directions have different magnitudes and therefore must be considered separately. Correlations have been presented for each in terms of Re. A method of presenting values of D_{e_r} and D_{e_L} is in terms of the Peclet number where $Pe = u d_p/D_e$. Measured values of P_{e_r} are in the range 5–11 and P_{e_L}, 1–2. The transfer of heat in packed beds is even more complicated than that of mass, since the solid phase can contribute to the transfer process as well as the gas. In the simplest model, which is the one considered here, the fluid and solid are imagined to act as a homogeneous mass in which heat is transferred by effective conduction. For the effective conductivity also, separate values must be specified for the radial and axial directions. These can be written as k_{e_r} and k_{e_L}. When measured experimentally it is found that k_{e_r} decreases markedly next to the wall, indicating an additional resistance there. This may be taken into account in deriving an average value of k_{e_r} or by introducing a wall heat transfer coefficient, α_w, where

$$\alpha_w(T_r - T_w) = -k_{e_r}\left(\frac{\partial T}{\partial r}\right).$$

k_{e_r} varies linearly with Re. Yagi and Kunii (1957) have built up a theory of heat transfer in packed beds according to which the effective conduction results from two contributions, static and dynamic. The static, measured in the absence of flow, considers heat transferred in the fluid and solid by conduction and radiation, and the dynamic is based on data on effective radial diffusivity. The effective thermal conductivity in an axial direction, k_{e_L}, is usually neglected since its contribution relative to that due to overall flow is very low.

The partial differential equations required to describe the mass and energy balances for reaction in a packed bed having significant radial contribution, will now be developed.

Two-Dimensional Model

Mass Balance. Suppose the reaction taking place is A → products. Consider a mass balance for species A over an increment of bed, volume $2\pi r\, dr\, dZ$, shown in Fig. 2.5.

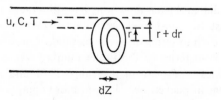

FIG. 2.5. Element of packed bed reactor

Mass in = mass out + disappearance by reaction

$$2\pi r dr u C \qquad\qquad - \qquad\qquad 2\pi r dZ \frac{D_{e_r}}{u}\left(\frac{\partial(uC)}{\partial r}\right)$$

(axially by mass flow) (radially by diffusion)

$$= 2\pi r dr\left\{uC+\left(\frac{\partial(uC)}{\partial Z}\right)dZ\right\} - 2\pi(r+dr)dZ\frac{D_{e_r}}{u}\left\{\left(\frac{\partial(uC)}{\partial r}\right)+\left(\frac{\partial^2(uC)}{\partial r^2}\right)dr\right\}$$

(axially by mass flow) (radially by diffusion

$$+2\pi r dr dZ(-r_c)\rho_{\mathrm{B}}. \tag{2.58}$$

(change by chemical reaction)

Multiplying out and rearranging we get

$$-2\pi r dr dZ\left(\frac{\partial(uC)}{\partial Z}\right)+2\pi r dZ\ \frac{D_{e_r}}{u}\left(\frac{\partial^2(uC)}{\partial r^2}\right)dr$$

$$+2\pi dr dZ\ \frac{D_{e_r}}{u}\left(\frac{\partial(uC)}{\partial r}\right)-2\pi r dr dZ(-r_c)\rho_{\mathrm{B}}=0$$

where second order differentials are neglected on expansion. Dividing throughout by $2\pi r dr dZ$, the result becomes

$$\frac{\partial(uC)}{\partial Z}-\frac{D_{e_r}}{u}\left\{\left(\frac{\partial^2(uC)}{\partial r^2}\right)+\frac{1}{r}\left(\frac{\partial(uC)}{\partial r}\right)\right\}+(-r_c)\rho_{\mathrm{B}}=0. \tag{2.59}$$

(For simplicity in presentation subscript A has been omitted, it being implied that symbols C and X refer to substance A).

Sometimes it is more convenient to work in terms of the fraction of reactant converted, X, where

$$X=\frac{u_0 C_0-uC}{u_0 C_0},$$

and $$d(uC)=-u_0 C_0\ dX.$$

Equation (2.59) is then written as

$$\frac{\partial X}{\partial Z}-\frac{D_{e_r}}{u}\left\{\left(\frac{\partial^2 X}{\partial r^2}\right)+\frac{1}{r}\left(\frac{\partial X}{\partial r}\right)\right\}-\frac{(-r_c)\rho_{\mathrm{B}}}{u_0 C_0}=0. \tag{2.60}$$

In the above derivation axial diffusion has been neglected and D_{e_r} has been written for the effective diffusion coefficient in the radial direction. Further, the linear velocity, u, has been considered constant over the tube cross-section. C refers to the concentration of A at any position in the tube and C_0 is the entrance concentration.

Heat Balance. Similarly a heat balance may be formulated for the incremental volume $2\pi r dr dZ$.

Heat in = heat out + heat change by chemical reaction

$$2\pi r dr \rho u C_p T_Z \qquad\qquad - \qquad 2\pi r dZ k_{e_r} \frac{\partial T}{\partial r}$$

(axially by flow)   (radially by conduction)

$$= 2\pi r dr \rho u C_p \left\{T_Z + \left(\frac{\partial T}{\partial Z}\right)dZ\right\} - 2\pi(r+dr)dZ\, k_{e_r}\left\{\left(\frac{\partial T}{\partial r}\right) + \left(\frac{\partial^2 T}{\partial r^2}\right) dr\right\}$$

(axially by flow)   (radially by conduction)

$$+ 2\pi r dr dZ(-r_c)\rho_B \Delta H. \tag{2.61}$$

(change by chemical reaction)

Multiplying out and rearranging gives

$$-2\pi r dr \rho u C_p\left(\frac{\partial T}{\partial Z}\right)dZ + 2\pi r dZ k_{e_r}\left(\frac{\partial^2 T}{\partial r^2}\right) dr + 2\pi dr dZ k_{e_r}\left(\frac{\partial T}{\partial r}\right)$$

$$-2\pi r dr dZ(-r_c)\rho_B \Delta H = 0.$$

Dividing by $2\pi r dr dZ$ gives

$$-\rho u C_p\left(\frac{\partial T}{\partial Z}\right) + k_{e_r}\left\{\left(\frac{\partial^2 T}{\partial r^2}\right) + \frac{1}{r}\left(\frac{\partial T}{\partial r}\right)\right\} - (-r_c)\,\rho_B \Delta H = 0$$

or

$$\frac{\partial T}{\partial Z} - \frac{k_{e_r}}{G C_p}\left\{\left(\frac{\partial^2 T}{\partial r^2}\right) + \frac{1}{r}\left(\frac{\partial T}{\partial r}\right)\right\} + \frac{(-r_c)\rho_B \Delta H}{G C_p} = 0 \tag{2.62}$$

where $G = \rho u$.

G is the molar flow rate/unit area (kmol/m^{2}s), ρ the density (kmol/ m^3) and u the linear velocity (m/s), or these may be expressed in mass units if $(-r_c)$ is in mass units. C_p is the gas stream heat capacity. k_{e_r} is the radial effective conductivity assumed constant over the cross-section.

If longitudinal diffusion is not neglected in relation to axial flow, the terms

$$-2\pi \frac{r dr D_{e_L}}{u} \left(\frac{\partial (uC)}{\partial Z}\right)$$

and

$$-2\pi \frac{rdr\,D_{e_L}}{u}\left\{\left(\frac{\partial(uC)}{\partial Z}\right)+\left(\frac{\partial^2(uC)}{\partial Z^2}\right)dZ\right\}$$

must be included on the L.H.S. and R.H.S. of equation (2.58), leading to the final equation

$$\frac{\partial X}{\partial Z}-\frac{D_{e_r}}{u}\left\{\left(\frac{\partial^2 X}{\partial r^2}\right)+\frac{1}{r}\left(\frac{\partial X}{\partial r}\right)\right\}-\frac{D_{e_L}}{u}\left(\frac{\partial^2 X}{\partial Z^2}\right)-\frac{(-r_c)\rho_B}{u_0 C_0}=0. \quad (2.63)$$

Similarly, if axial conduction is considered to be important, the terms

$$-2\pi r dr k_{e_L}\left(\frac{\partial T}{\partial Z}\right)$$

and

$$-2\pi r dr k_{e_L}\left\{\left(\frac{\partial T}{\partial Z}\right)+\left(\frac{\partial^2 T}{\partial Z^2}\right)dZ\right\}$$

must be included in equation (2.61), giving

$$\frac{\partial T}{\partial Z}-\frac{k_{e_r}}{GC_p}\left\{\left(\frac{\partial^2 T}{\partial r^2}\right)+\frac{1}{r}\left(\frac{\partial T}{\partial r}\right)\right\}-\frac{k_{e_L}}{GC_p}\left(\frac{\partial^2 T}{\partial Z^2}\right)+\frac{(-r_c)\rho_B\Delta H}{GC_p}=0. \quad (2.64)$$

The method of solution of these second order partial differential equations of the boundary value type, is through replacement of the differential equations by finite difference equations.

Equations (2.60) and (2.62) are then transformed into the forms:

$$\frac{X_{m,n+1}-X_{m,n}}{\Delta Z}-\frac{D_{e_r}}{u}\left[\frac{X_{m+1,n}-X_{m,n}-(X_{m,n}-X_{m-1,n})}{(\Delta r)^2}\right.$$
$$\left.+\frac{X_{m+1,n}-X_{m,n}}{m(\Delta r)^2}\right]-\frac{(-r_c)_{av.}\rho_B}{u_0 C_0}=0 \qquad (2.65)$$

and

$$\frac{T_{m,n+1}-T_{m,n}}{\Delta Z}-\frac{k_{e_r}}{GC_p}\left[\frac{T_{m+1,n}-T_{m,n}-(T_{m,n}-T_{m-1,n})}{(\Delta r)^2}\right.$$
$$\left.+\frac{T_{m+1,n}-T_{m,n}}{m(\Delta r)^2}\right]+\frac{(-r_c)_{av.}\rho_B\Delta H}{GC_p}=0 \qquad (2.66)$$

where $r=m\Delta r$ and $Z=n\Delta Z$, m and n being the number of increments in the radial and axial directions respectively. These equations

c

can be rearranged to give

$$X_{m,n+1} = X_{m,n} + A\left[\frac{1}{m}(X_{m+1,n} - X_{m,n}) + X_{m+1,n}\right.$$

$$\left. -2X_{m,n} + X_{m-1,n}\right] + M(-r_c)_{\text{av.}}$$

where $A = \dfrac{\Delta Z D_{e_r}}{(\Delta r)^2 u}$ and $M = \dfrac{\rho_B \Delta Z}{u_0 C_0}$,   (2.67)

and $T_{m,n+1} = T_{m,n} + B\left[\frac{1}{m}(T_{m+1,n} - T_{m,n}) + T_{m+1,n}\right.$

$$\left. -2T_{m,n} + T_{m-1,n}\right] - D(-r_c)_{\text{av.}}$$

where $B = \dfrac{\Delta Z k_{e_r}}{(\Delta r)^2 G C_p}$ and $D = \dfrac{\rho_B \Delta H \Delta Z}{G C_p}$.   (2.68)

From these equations the conversion and temperature can be computed at any point in the bed from known entrance composition and temperature distributions, $X_{m,0}$ and $T_{m,0}$. The procedure is to calculate $X_{m,1}$ and $T_{m,1}$ from the initial values at all values of radial increments determined by the chosen size of ΔZ and Δr. Values of D_{e_r}, u, G, C_p, k_{e_r} and ρ_B must be available together with rate data, $(-r_c)$, and values for heat of reaction, ΔH.

A difficulty arises in application of the equations to the case when $m = 0$, i.e. at the centre of the bed. This is because the terms $[(1/r)(\partial X/\partial r)]$ and $[(1/r)(\partial T/\partial r)]$ are indeterminate at the centre, since r and the differentials all approach zero. If the numerators, $(\partial X/\partial r)$ and $(\partial T/\partial r)$, and the denominator, r, are differentiated separately, we get $(\partial^2 X/\partial r^2)$ and $(\partial^2 T/\partial r^2)$. Hence equations (2.60) and (2.62) become

$$\frac{\partial X}{\partial Z} = \frac{2D_{e_r}}{u}\left(\frac{\partial^2 X}{\partial r^2}\right) + \frac{(-r_c)\rho_B}{u_0 C_0}   (2.69)$$

and $$\frac{\partial T}{\partial Z} = \frac{2k_{e_r}}{G C_p}\left(\frac{\partial^2 T}{\partial r^2}\right) - \frac{(-r_c)\rho_B \Delta H}{G C_p},   (2.70)$$

and the difference forms are written

$$X_{0,n+1} = X_{0,n}+2A[(X_{1,n}-X_{0,n})-(X_{0,n}-X_{-1,n})]+M(-r_c)_{av.}$$
(2.71)

and

$$T_{0,n+1} = T_{0,n}+2B[(T_{1,n}-T_{0,n})-(T_{0,n}-T_{-1,n})]-D(-r_c)_{av.}$$
(2.72)

By symmetry $T_{-1,n} = T_{1,n}$ and $X_{-1,n} = X_{1,n}$ and therefore, from (2.71) and (2.72),

$$X_{0,n+1} = X_{0,n}+2A(2X_{1,n}-2X_{0,n})+M(-r_c)_{av.} \qquad (2.73)$$

$$\text{and } T_{0,n+1} = T_{0,n}+2B(2T_{1,n}-2T_{0,n})-D(-r_c)_{av.} \qquad (2.74)$$

The method of solution is by trial and error. The rate at the end of the axial increment is assumed for each radial position, and the average rate calculated from this value and the known initial rate of reaction. From equation (2.67), or equation (2.73) if the centre tube position is being considered, $X_{m,1}$ is calculated from $X_{m,0}$ and from equation (2.68), or equation (2.74) for the centre line case, $T_{m,1}$ is calculated. The rate at the end of the increment can then be calculated from the computed values of $T_{m,1}$ and $X_{m,1}$ and an average rate calculated. Agreement should be obtained between this value and the original assumed $(-r_c)_{av}$ before proceeding to the next axial increment.

Some comment must be made regarding the calculation at the wall since in equation (2.65), $X_{m+1,n}$ would in fact refer to a position outside the reactor wall. The concentration gradient at the wall must be zero and it is reasonable to suppose the conversion symmetrical about the wall, which means the $X_{m+1,n}$ can be taken equal to $X_{m-1,n}$. In the case of temperature, the wall temperature will be fixed and it is therefore not necessary to apply the temperature equation at the wall.

In making calculation for one depth increment from the earlier one, it is desirable to 'smooth' the calculated values and use the smoothed values in the next calculation.

The desired result in such a calculation is the mean conversion of material leaving the reactor, together with information on the temperature profile in the reactor.

The moles converted in an element dr will be given by $XG2\pi r\,dr$, where G is molar flow rate/unit area. Integrating over all radial elements

$$\pi r_0^2 X_{me} G = \int_0^{r_0} XG2\pi rdr$$

$$\text{or } X_{me} = \frac{2\int_0^{r_0} Xrdr}{r_0^2}. \tag{2.75}$$

Replacing r/r_0 by q gives

$$X_{me} = 2\int_0^1 Xqdq. \tag{2.76}$$

Hence a plot of Xq versus q allows X_{me} to be calculated from the area under the graph.

The application of the one-dimensional and two-dimensional models to an exothermic catalytic reaction is considered in Chapter 7.

REFERENCES

COOPER, A. R., and JEFFREYS, G. V. 1971. Chemical kinetics and reactor design. Oliver and Boyd, Edinburgh.

LEVENSPIEL, O. 1962. Chemical reaction engineering. Wiley, New York.

SMITH, J. M. 1970. Chemical engineering kinetics, 2nd Ed. McGraw-Hill, New York.

YAGI, S., and KUNII, D. 1957. *A.I.Ch.E.J.*, **3**, 373.

3

REVERSIBLE EXOTHERMIC REACTIONS

3.1 Introduction

A very important class of industrial reactions is described by the terms 'reversible' and 'exothermic'. The distinguishing features are the evolution of heat during reaction, the attainment of a state of equilibrium in which conversion to product is less than complete and a decrease in product yield at equilibrium as the reaction temperature is raised. Processes of industrial importance which fall into this class include the synthesis of ammonia from nitrogen and hydrogen, sulphur dioxide oxidation and methanol synthesis.

Before considering some of the features influencing the design and conditions of operation of reactors for this class of industrially important reversible exothermic reactions, the case of a simple first order reversible reaction will be examined in some detail, since the simple case is easier to analyse in a quantitative fashion. The conclusions drawn from the simple reaction analysis will be seen to apply qualitatively to the more complex industrial reactions.

3.2 Reversible First Order Reaction

A reversible first order reaction, $A \rightleftharpoons B$, will be considered. $\Delta G°_{298°K}$ for the reaction is -10000 J/mol and $\Delta H°$ is assumed constant at -40000 J/mol over the temperature range 273–573°K. The rate constant for the forward reaction can be calculated from the Arrhenius expression $k_1 = 10^9 \, e^{-60000/RT}$ where the units of k_1 are min^{-1} and the energy of activation is 60000 J/mol.

3.2.1 EFFECT OF TEMPERATURE ON THE EQUILIBRIUM YIELD

In this example, where $\Delta H°$ remains constant in the temperature range under consideration, the integrated form of the van't Hoff equation, equation (1.41), may be employed to calculate the value of the equilibrium constant at any temperature if a value at a single temperature is known.

TABLE 3.1

Equilibrium constant and conversion as a function of temperature

Temperature, °K	$(1/T) \times 10^3$ $(°K)^{-1}$	$\log K$	K	X_{A_e}
273	3·663	2·3938	247·7	0·996
298	3·356	1·7526	56·57	0·983
313	3·195	1·4162	26·07	0·963
323	3·096	1·2093	16·19	0·942
333	3·003	1·0150	10·35	0·912
338	2·959	0·9231	8·377	0·893
343	2·915	0·8312	6·779	0·871
348	2·874	0·7455	5·565	0·848
353	2·833	0·6598	4·568	0·820
358	2·793	0·5763	3·770	0·790
363	2·755	0·4970	3·141	0·758
368	2·717	0·4176	2·616	0·723
373	2·681	0·3424	2·200	0·687
378	2·646	0·2692	1·859	0·650
383	2·611	0·1961	1·570	0·611
388	2·577	0·1251	1·334	0·572
393	2·545	0·0583	1·144	0·534
398	2·513	−0·0086	0·980	0·495
403	2·481	−0·0755	0·840	0·457
408	2·451	−0·1381	0·728	0·421
413	2·421	−0·2008	0·630	0·386
418	2·392	−0·2614	0·548	0·354
423	2·364	−0·3199	0·479	0·324
433	2·309	−0·4348	0·367	0·269
473	2·114	−0·8421	0·144	0·126
523	1·912	−1·2641	0·054	0·052

The equilibrium constant, K, is evaluated at 298°K using equation (1.34):

$$\log K_{298°K} = \frac{10000}{2·303 \times 8·314 \times 298} = 1·753$$

and $K_{298°K} = 56·57$.

Equation (1.41) is applied, either analytically or in a graphical form, to obtain K at any other temperature.

For example at $T = 373°K$

$$\log \frac{K}{56·57} = \frac{40000}{2·303 \times 8·314} \left(\frac{1}{373} - \frac{1}{298} \right)$$

and $K_{373°K} = 2·20$.

Values of K at other temperatures are shown in Table 3.1 or may be deduced from Fig. 3.1 which is a plot of log K against $1/T$ $(°K)^{-1}$.

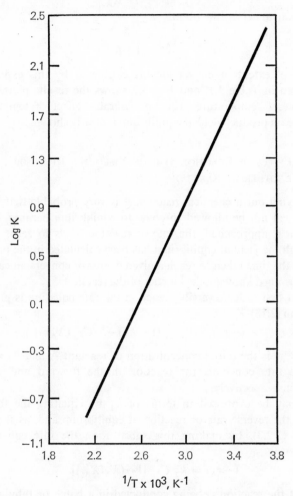

FIG. 3.1. Temperature dependence of equilibrium constant
(data in Table 3.1)

Suppose that the extent of conversion is given by X_{A_e} per mole of initial reactant A. At equilibrium there will be $1 - X_{A_e}$ moles of A and X_{A_e} moles of B, and from equation (1.55) the equilibrium equation will be given by

$$K = \frac{X_{A_e}}{1 - X_{A_e}}. \tag{3.1}$$

On rearrangement we obtain

$$X_{A_e} = \frac{K}{1 + K}. \tag{3.2}$$

Values of extent of conversion, as calculated by this expression, are given in Table 3.1, and Fig. 3.2 shows the results plotted as a function of temperature. The unfavourable effect of temperature increase on product yield at equilibrium is clearly shown.

3.2.2   EFFECT OF REACTION TIME ON YIELD IN A REVERSIBLE EXOTHERMIC REACTION

In carrying out a chemical reaction it is very probable that the reaction will not be allowed to come to equilibrium because, as this position is approached, the rate of reaction tends to zero. Indeed although the yield at equilibrium has been calculated in the previous section the time taken to reach a given degree of conversion can only be determined knowing the kinetics of the reaction.

For a first order reversible reaction the rate equation is given by equation (2.18)

$$(-r_A) = k_1 C_{A_0}(1 - X_A) - k_2 C_{A_0}(X_A) \tag{2.18}$$

where C_{A_0} is the initial concentration of reactant A and k_1 and k_2 are the rate constants for reaction in the forward and reverse directions respectively.

k_2 may be expressed in terms of k_1 by equating the forward rate to the reverse rate of reaction at equilibrium and, as shown in section (2.2.3), by making this substitution the rate equation is rewritten as

$$(-r_A) = k_1 C_{A_0}[1 - (X_A/X_{A_e})]. \tag{2.19}$$

Now, if the reaction is being conducted in a batch or tubular plug flow reactor in which the temperature, pressure and number of moles of reacting mixture are everywhere the same, the holding time in the reactor will be given by equation (2.36)

$$\tau = t = C_{A_0} \int_0^{X_A} \frac{dX_A}{(-r_A)}.$$

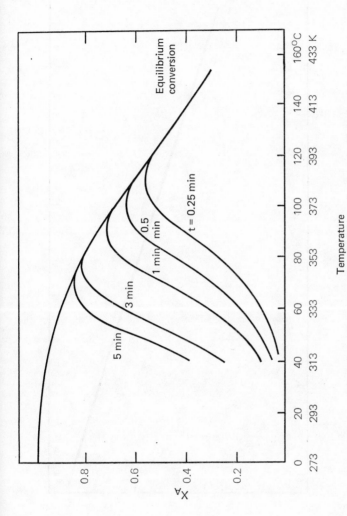

Fig. 3.2. Conversion as a function of temperature and holding time in the reactor

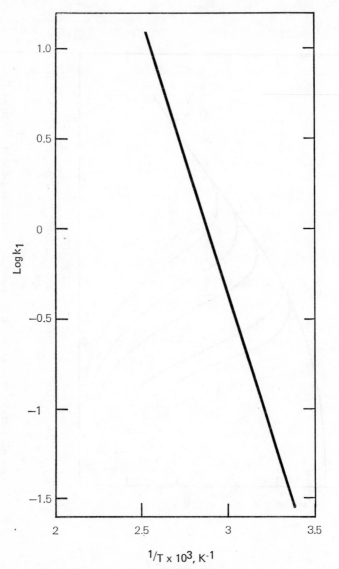

FIG. 3.3. Temperature dependence of rate constant for forward reaction

Then substituting for $(-r_A)$ using equation (2.19)

$$t = \frac{1}{k_1} \int_0^{X_A} dX_A \Big/ \left(1 - \frac{X_A}{X_{A_e}}\right).$$

On integrating this expression we obtain

$$k_1 t = -X_{A_e} \ln[1 - (X_A/X_{A_e})]$$

or $\qquad X_A = X_{A_e}(1 - e^{-k_1 t/X_{A_e}}).$ \hfill (2.22)

Equation (2.22) may be used to calculate the conversion of A resulting from reaction in a given period of time at any temperature T, or to find the time required for a given extent of conversion at any temperature T.

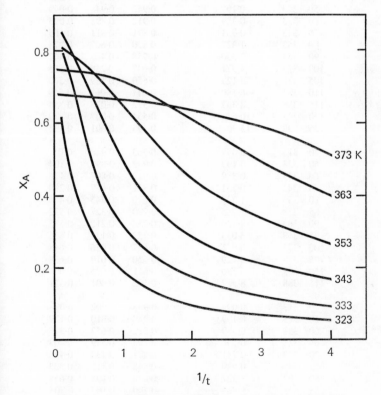

FIG. 3.4. Conversion as a function of temperature and reciprocal of time spent in the reactor

For the reaction being considered, and using the previously determined equilibrium conversion values, X_{A_e}, and the values of k_1 calculated from the given Arrhenius expression (plot of log k_1 against $1/T$ appears in Fig. 3.3), values of X_A have been calculated using equation (2.22) over a range of assumed residence times.

TABLE 3.2

Conversion as a function of temperature and time spent in the reactor

Temperature °C	°K	k_1 min^{-1}	t min	X_{A_e}	$e^{-k_1 t/X_{Ae}}$	X_A
40	313	0·099	0·25	0·963	0·974	0·025
50	323	0·199		0·942	0·948	0·049
60	333	0·389		0·912	0·898	0·093
70	343	0·731		0·871	0·811	0·165
80	353	1·327		0·820	0·667	0·273
90	363	2·330		0·758	0·463	0·407
100	373	3·971		0·687	0·236	0·525
105	378	5·129		0·650	0·139	0·560
110	383	6·580		0·611	0·068	0·569
115	388	8·387		0·572	0·025	0·558
120	393	10·63		0·534	0·007	0·530
125	398	13·39		0·495	0·001	0·495
40	313	0·099	0·5	0·963	0·948	0·05
50	323	0·199		0·942	0·890	0·104
60	333	0·389		0·912	0·807	0·176
70	343	0·731		0·871	0·657	0·299
80	353	1·327		0·820	0·445	0·455
85	358	1·765		0·790	0·327	0·532
90	363	2·330		0·758	0·215	0·595
95	368	3·053		0·723	0·121	0·635
100	373	3·971		0·687	0·056	0·648
105	378	5·129		0·650	0·019	0·638
110	383	6·580		0·611	0·005	0·608
115	388	8·387		0·572	0·001	0·571
40	313	0·099	1	0·983	0·902	0·098
50	323	0·199		0·942	0·810	0·179
60	333	0·389		0·912	0·652	0·317
65	338	0·536		0·893	0·549	0·403
70	343	0·731		0·871	0·432	0·495
75	348	0·989		0·848	0·312	0·583
80	353	1·327		0·820	0·198	0·658
85	358	1·765		0·790	0·107	0·705
90	363	2·330		0·758	0·046	0·723

TABLE 3.2 (*continued*)

Temperature		k_1	t	X_{Ae}	$e^{-k_1 t/X_{Ae}}$	X_A
°C	°K	min^{-1}	min			
95	368	3·053		0·723	0·015	0·712
100	343	3·971		0·687	0·003	0·685
40	313	0·099	3	0·963	0·735	0·255
50	323	0·199		0·942	0·530	0·443
60	333	0·389		0·912	0·278	0·658
65	338	0·536		0·893	0·165	0·746
70	343	0·731		0·871	0·081	0·800
75	348	0·989		0·848	0·030	0·823
80	353	1·327		0·820	0·008	0·813
85	358	1·765		0·790	0·001	0·789
40	313	0·099	5	0·963	0·598	0·387
50	323	0·199		0·942	0·348	0·614
60	333	0·389		0·912	0·118	0·804
65	338	0·536		0·893	0·050	0·848
70	343	0·731		0·871	0·015	0·858
75	348	0·989		0·848	0·003	0·845
80	353	1·327		0·820	0·001	0·820

The calculated values are presented in Table 3.2 and superimposed on Fig. 3.2. An alternative plot which is often useful is shown in Fig. 3.4. Here conversion values are plotted against (time)$^{-1}$; this form of plot is often used in presenting data for industrial reactions where conversion is plotted against the space velocity (defined in 2.3.1.1).

It is apparent from Fig. 3.2 that for each reaction time there is a maximum conversion, the value of which is higher the longer the reaction time. Also, the longer the reaction time, the lower is the temperature at which the maximum occurs. Thus in contrast to the situation for irreversible reactions or reversible endothermic reactions, the extent of conversion does not invariably increase as the temperature increases. This is because the reverse reaction becomes increasingly important as the temperature increases.

In a chemical process it is not generally a maximum conversion that is the primary requirement but rather that the maximum rate of production should be obtained, and Fig. 3.2 may be examined to see what effect change in reaction time has on production rate. For reaction in a constant volume flow reactor the residence time is inversely proportional to the molar flow rate of reactant. At low temperatures it may be seen that conversion is not directly proportional to reaction time; in fact proportionally higher conversions are obtained the shorter the residence time.

For example, at 333°K the following results are deduced:

t, min	X_A	X_A/t
0·25	0·093	0·372
0·5	0·176	0·352
1·0	0·317	0·317
3·0	0·658	0·219
5·0	0·804	0·161

Since residence time is inversely proportional to flowrate this implies that production rates will be greater at high flow rates than at low.

For example, suppose that at $t = 0.5$ min, the flow rate is given by F_{Ao}, then at $t = 1$ min, the flow rate is $0.5 F_{Ao}$. Hence at $t = 0.5$ min, $F_{Ao}X_A = 0.176 F_{Ao}$, and at $t = 1$ min, $F_{Ao}X_A = (0.317 \times 0.5)F_{Ao} = 0.158 F_{Ao}$.

It is deduced therefore that for maximum production rates, reactant flow rates as high as possible should be used. Of course for overall optimisation of cost of production other factors such as cost of pumping, pressure drop consideration etc. must be taken into account. If it were supposed that for the given reactor a reactant flowrate corresponding to a residence time of 0·25 min was the maximum value which could be used in the case under discussion, then, from Fig. 3.2, the optimum isothermal production rate would be obtained at a temperature of 383°K when the conversion would be 0·569.

3.2.3   Optimisation of Reaction Rates in a Reversible Exothermic Reaction

In the previous section it was deduced that production rates will be increased by operating at a high flow rate and an associated high isothermal temperature. However, it might be possible to achieve still higher outputs if temperature optimisation could be employed rather than operating at a fixed temperature.

Suppose the aim is to achieve the same conversion as obtained under the 'best' isothermal conditions deduced in the previous section i.e. $X_A = 0.569$, obtained at a temperature of 383°K. It is required to calculate the reduction in size of reactor possible for the same conversion, if the reactant flow rate and initial reactant concentration are the same as when operating under optimum temperature conditions. For a tubular plug flow reactor

$$\tau = \frac{VC_{A0}}{F_{A0}} = C_{A0}\int_0^{X_A} \frac{dX_A}{(-r_A)}. \tag{2.36}$$

In essence the value of V is to be calculated under the condition that the integral on the R.H.S. has a minimum value for a conversion of 56.9%. This will be achieved by ensuring that the reaction rate is at its maximum possible value at all stages of reaction, i.e. that

$$\frac{d(-r_A)}{dT} = 0.$$

The value of V obtained will be the minimum possible for the given feed rate, feed concentration and output required. Now for a first order reversible reaction

$$(-r_A) = k_1 C_{A0} (1 - X_A) - k_2 C_{A0} (X_A) \tag{2.18}$$

$$= A_1 \exp(-E_1/RT)C_{A0}(1 - X_A) - A_2 \exp(-E_2/RT)C_{A0}(X_A). \tag{3.3}$$

By differentiating this expression with respect to temperature and setting the result equal to zero we obtain

$$\frac{d(-r_A)}{dT} = A_1 \exp(E_1/RT)\frac{E_1}{RT^2}C_{A0}(1 - X_A)$$

$$- A_2 \exp(-E_2/RT)\frac{E_2}{RT^2}C_{A0}(X_A) = 0$$

$$= k_1 \frac{E_1}{RT^2} C_{A0} (1 - X_A) - k_2 \frac{E_2}{RT^2} C_{A0} (X_A) = 0.$$

From this equation we deduce that

$$K = \frac{k_1}{k_2} = \frac{X_A}{1 - X_A} \cdot \frac{E_2}{E_1}. \tag{3.5}$$

For each assumed extent of conversion, X_A, and knowing the energies of activation for the forward and back reactions a value of the equilibrium constant, K, may be calculated. The temperature associated with this constant can then be inferred from Fig. 3.1. This temperature, (T_{max}), is that at which the reaction rate is at a maximum for the given conversion. The maximum rate is calculated by applying equation (2.18) at T_{max}, k_1 and k_2 being computed at this temperature

$$(-r_A)_{max} = k_{1(T_{max})}C_{A0}(1-X_A) - k_{2(T_{max})}C_{A0}(X_A). \qquad (2.18)$$

The reaction time necessary in order to achieve the required conversion, X_A, can be obtained by carrying out a graphical integration of $1/(-r_A)_{max}$ against X_A. It can be seen from equation (3.5), that as X_A increases, K increases which implies that as the reaction proceeds the optimum temperature of operation for maximum rate should decrease, since the value of K decreases with increase in temperature.

Thus, in order to attain maximum production rate for a reversible exothermic reaction, a falling temperature sequence is required, i.e. the reaction temperature should drop down the length of the reactor in the case of a flow reactor or decrease with time of reaction in the case of a batch reactor. In the early stages of reaction the reverse reaction is of little importance because of the low product concentration. The reaction rate can then be as high as possible, which implies the use of a high temperature, but, as the reaction proceeds, the product concentration builds up, and in order to maintain the rate at a maximum the temperature must fall, thereby bringing the reaction further away from the equilibrium point. It should be observed however from equation (3.5) that as $X_A \to 0$, K also $\to 0$ which implies that $T_{max} \to \infty$. Obviously from a practical point of view this condition must be avoided and an upper limit to the maximum allowable temperature must be set.

In Table 3.3 data are presented for the maximum rate of reaction in the case under consideration at various values of X_A and for an assumed initial reactant concentration of 1 kmol/m^3. Values of $1/(-r_A)_{max}$ are plotted against X_A in Fig. 3.5 and the value of τ deduced. A maximum allowable temperature of 423°K is assumed.

From the area under the curve in Fig. 3.5 a value of $\tau = 0\cdot16$ min is obtained. Since, for the same conversion, the time spent in the isothermal reactor was 0·25 min, and since the reactant flow rate and initial concentration are to be the same in the two cases, the reactor volume for the temperature optimised case can be reduced by a

TABLE 3.3

Maximum reaction rate as a function of conversion and associated reaction temperature

X_A	T_{max} °C	°K	$(-r_A)_{max}$ kmol/m³ min		$1/(-r_A)_{max}$
0	(∞)	(∞)	39·07	(423°K)	0·026
0·1	(199)	(472)	27·00	(423°K)	0·037
0·2	(157)	(430)	14·93	(423°K)	0·067
0·3	137	410	6·371		0·157
0·4	122	395	2·798		0·357
0·45	115	388	1·848		0·541
0·5	101·5	374·5	1·252		0·798
0·6	97	370	0·543		1·842
0·65	91	364	0·344		2·907
0·7	85	358	0·212		4·717
0·8	71	344	0·061		16·393

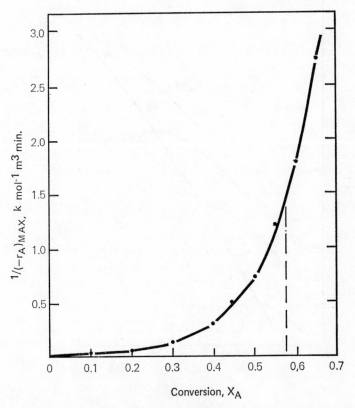

FIG. 3.5. Relation between conversion and reciprocal of maximum rate

factor of $[(0.25-0.16)/0.16] = 56.15\%$ and still maintain the same conversion.

Alternative methods of determining the optimum temperature sequence may be employed. For example in the case of a complex rate equation, differentiation and setting $d(-r_A)/dT = 0$ may not be easy in practice. In such cases it may be simpler to evaluate the reaction rate at values of X_A over a range of temperatures; the locus of maximum rate points for each conversion can then be drawn. Such a plot is illustrated for the present case in Fig. 3.6. The agreement in $(-r_A)_{max}$, as calculated analytically and as plotted in Fig. 3.6, is to be noted.

Yet another method which applies to first order reversible

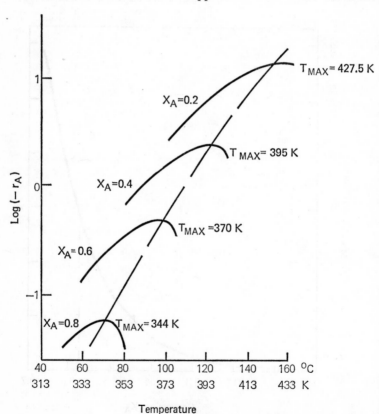

FIG. 3.6. Reaction rate as a function of temperature and conversion

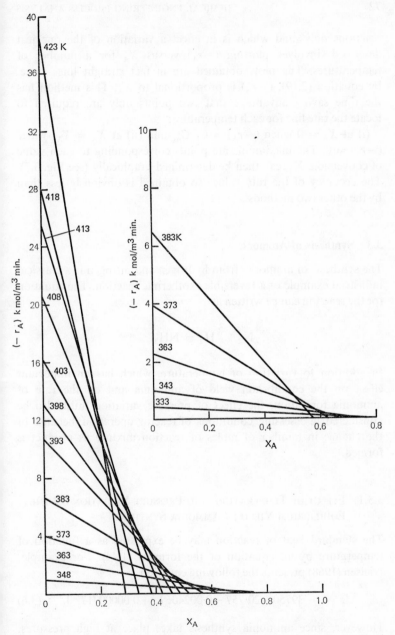

FIG. 3.7. Rate as a function of conversion and temperature
for a first-order reversible reaction

reactions only, and which is in effect a variation of the one just described, involves plotting $(-r_A)$ versus X_A for a number of temperatures. The plots obtained are in fact straight lines since, by equation (2.19), $(-r_A)$ is proportional to X_A. This method has the time saving advantage that two points only are required to locate the rate line for each temperature:

(i) at $X_A = 0$ when $(-r_A) = k_1 C_{A_0}$ and (ii) at $X_A = X_{A_e}$ when $(-r_A) = 0$. The maximum rate points corresponding to each value of conversion, X_A, can then be determined graphically (see Fig. 3.7). The accuracy of the rate values so obtained is obviously less than by the other two methods.

3.3 Synthesis of Ammonia

The synthesis of ammonia from hydrogen and nitrogen is a classical industrial example of a reversible exothermic reaction. The equation for the reaction can be written as

$$\frac{1}{2}N_2 + \frac{3}{2}H_2 \rightleftharpoons NH_3.$$

In addition to variation of temperature, which has an important effect on the equilibrium yield of ammonia and on the rate of ammonia formation, the effect of pressure variation must also be considered in assessing conditions of reactor operation, because of the change in number of moles of reaction mixture as product is formed.

3.3.1 EFFECT OF TEMPERATURE AND PRESSURE VARIATION ON THE EQUILIBRIUM YIELD IN AMMONIA SYNTHESIS

The standard heat of reaction may be expressed as a function of temperature by an equation of the form of (1.44). For example, Nielsen (1968) presents the following equation

$$\Delta H^\circ = -39750 - 20 \cdot 75T - 0 \cdot 002406T^2 + 0 \cdot 0000071T^3 \, \text{J}. \quad (3.6)$$

However, since ammonia synthesis takes place at high pressures, pressure as well as temperature will be a factor determining the heat of reaction. Gillespie and Beattie (1930) used the following equation to arrive at a value for ΔH

$$\Delta H = -\left[2\cdot281+\frac{351\cdot8}{T}+\frac{1924\times10^6}{T^3}\right]P$$
$$-22\cdot38T-1\cdot056\times10^{-3}T^2+7\cdot079\times10^{-6}T^3$$
$$-38310 \text{ J}. \tag{3.7}$$

The reaction is of course exothermic and at 873°K and 1 atm. pressure

$$\Delta H^0 = -54974 \text{ J/mol}$$

by equation (3.6).

Substituting an equation for $\Delta H°$, derived from combination of free energy functions, in (1.46), Harrison and Kobe (1953) arrived at an expression relating the equilibrium constant to temperature in the range 500–1300°K:

$$\log K = 2250\cdot322T^{-1}-0\cdot85340-1\cdot51049 \log T$$
$$-2\cdot58987\times10^{-4}T+1\cdot48961\times10^{-7}T^2. \tag{3.8}$$

Values of the equilibrium constant over the useful temperature range are shown in Table 3.4.

TABLE 3.4

Ammonia synthesis: equilibrium constant as a function of temperature*

Temp., °K	log K	K	Temp., °K	log K	K
300	2·7825	606	900	−2·9278	$1\cdot181\times10^{-3}$
400	0·7504	5·628	1000	−3·2448	$5\cdot692\times10^{-4}$
500	−0·5218	$3\cdot008\times10^{-1}$	1100	−3·5058	$3\cdot121\times10^{-4}$
600	−1·3994	$3\cdot987\times10^{-2}$	1200	−3·7248	$1\cdot884\times10^{-4}$
700	−2·0434	$9\cdot048\times10^{-3}$	1300	−3·9109	$1\cdot229\times10^{-4}$
800	−2·5365	$2\cdot908\times10^{-3}$	1400	−4·0720	$8\cdot473\times10^{-5}$

* From Harrison and Kobe (1953).

Using equation (1.54) with the values of K from the above table, the percentage of ammonia at equilibrium at various temperatures and pressures can be evaluated assuming that the gas mixture is ideal. It was seen in section 1.6.1. that the assumption of an ideal gas mixture holds reasonably well up to a pressure of around 300 atm. for this reaction. Applied to the ammonia synthesis, equation (1.54) has the form

$$K'_f = \frac{x_{NH_3}}{(x_{N_2})^{1/2}(x_{H_2})^{3/2}} \frac{\gamma'_{NH_3}}{(\gamma'_{N_2})^{1/2}(\gamma'_{H_2})^{3/2}} \frac{1}{P}. \qquad (3.9)$$

The fugacity coefficients are calculated according to the method given in section 1.6.1.

Table 3.5 presents some experimental data of Larson and Dodge (1923, 1924) for the ammonia equilibrium. These are plotted in Fig. 3.8.

TABLE 3.5

Mole % NH_3 at equilibrium for a pure 3:1, $H_2:N_2$ mixture

Temp. (°K)	Pressure (atm.)			
	50	100	300	600
473	74·38	81·54	89·94	95·37
523	56·33	67·24	81·38	90·66
573	39·44	52·04	70·96	84·21
623	25·23	37·35	59·12	75·62
673	15·27	25·12	47·00	65·20
723	9·15	16·43	35·82	53·71
773	5·56	10·61	26·44	42·15
823	3·45	6·82	19·13	31·63

From Table 3.5 it is apparent that high equilibrium yields of ammonia are favoured by conditions of low temperature and high pressure. This deduction is of course based on thermodynamic considerations alone and it is necessary that the rates of ammonia formation under variable temperature and pressure conditions should also be considered.

3.3.2 RATE OF THE AMMONIA SYNTHESIS REACTION

Ammonia synthesis from nitrogen and hydrogen is a reversible reaction but the order of reaction in either direction is not simple and even today the mechanism of the reaction is not fully understood. The equation which best describes the rate is the Temkin (1940) equation which in its simplest form is given by

$$r = k_1 \frac{p_{N_2}(p_{H_2})^{1.5}}{p_{NH_3}} - k_2 \frac{p_{NH_3}}{(p_{H_2})^{1.5}}. \qquad (3.10)$$

More refined versions of this equation give a more accurate representation of the experimental results (Temkin *et al.*, 1963). For example, since we are dealing with a high pressure, high temperature reaction, it is better to replace partial pressures by fugacities (Dyson and Simon, 1968). However for the purpose of argument equation (3.10) is suitable for consideration. The resemblance of its form to that of equation (2.18) is apparent. k_1 is the rate constant for the forward or formation reaction and k_2 for the back or decomposition reaction. The rate of formation of ammonia increases as the partial pressures of hydrogen and nitrogen increase but decreases

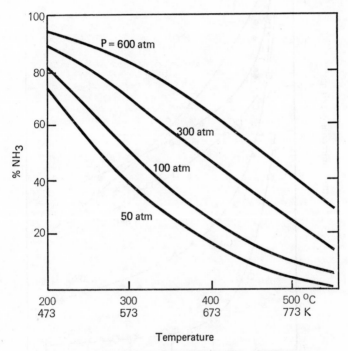

Fig. 3.8. Mole percentage of ammonia at equilibrium for a 3:1 = $H_2:N_2$ mixture

as ammonia is formed. The reaction rate is zero when equilibrium is reached.

The synthesis of ammonia takes place in a fixed bed catalytic reactor. The nature of the catalyst and some aspects of the mechanism will be considered in section 3.3.5. Valuable information on the ammonia synthesis reaction has been reported by Adams and Comings (1953) and by Nielsen (1968). Fig. 3.9 shows the percentage of ammonia at the exit of a reactor for a 3:1, $H_2:N_2$ mixture at 300 atm. as a function of space velocity at various temperatures. Space velocity is a more convenient concept than the reciprocal of the residence time used in section 3.2.2 for reactions in which

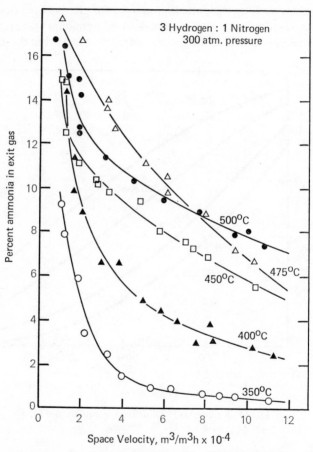

FIG. 3.9. Percentage ammonia in exit gas at 300 atm. pressure (Reprinted with permission from *Chem. Eng. Prog.* **49**, 359 (1953))

the number of moles varies with extent of reaction as in the present case. It is defined as the volume of reaction mixture, measured at 273°K and 1 atm. pressure, flowing per volume of catalyst per hour. The data are plotted in an alternative way in Fig. 3.10. The resemblance of the shapes of these curves to those calculated for a first order reversible reaction, and represented in Figs. 3.4. and 3.2, should be noted. In Fig. 3.9 the cross-over point involving the curves at 475°C and 500°C is explained by the fact that at low space velocities

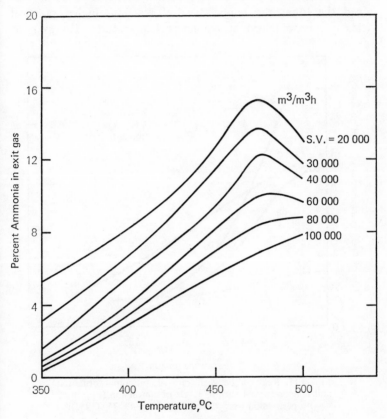

FIG. 3.10. Conversion as a function of temperature and space velocity

at these temperatures the conversion will be influenced overwhelmingly by the equilibrium yield, which is higher at 475°C than at 500°C, while at high space velocities the conversion will be controlled by reaction rate considerations. The data shown in Fig. 3.9, which are experimental, could have been calculated by substituting an accurate form of the Temkin rate equation into an equation of the form of (2.36). However when the reaction is complex the data are best obtained experimentally.

From Fig. 3.10 it is clear that increasing the temperature leads to an increase in yield of ammonia as long as the yield is not approaching the equilibrium yield. Since higher equilibrium yields are obtained the lower the temperature of reaction, higher practical yields can be obtained at the lower temperatures. However, as

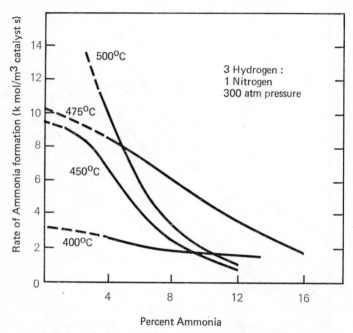

FIG. 3.11. Relation between rate of ammonia formation and temperature at various levels of conversion (Reprinted with permission from *Chem. Eng. Prog.* **49**, 359 (1953))

pointed out in section 3.2.2, the important factor is not that the yield should be high but that the specific rate of ammonia formation should be as high as possible. In industrial processing the parameter to be maximised is the space time yield, which is given by moles product at reactor exit/mole of entering feed times space velocity. The output of the reactor will then be (S.T.Y.) (reactor volume). Nielsen (1968) has pointed out that in ammonia synthesis space time yield increases with increasing space velocity because conversion is proportionally higher at high velocities than at low. This observation is again in accordance with the findings for a reversible first order reaction. Thus high space velocities are desirable in industrial plants. However there are limitations placed on the magnitude of space

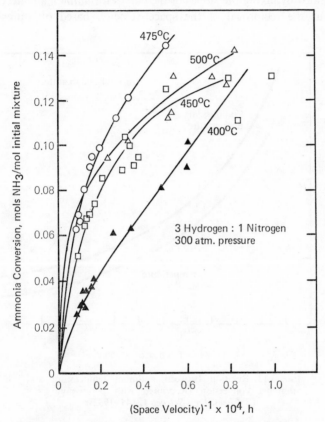

FIG. 3.12. Relation between fractional conversion to ammonia and reciprocal of space velocity (Reprinted with permission from *Chem. Eng. Prog.* **49**, 359 (1953))

velocities by the pressure drop through the catalyst bed, the difficulty of heat transfer from the catalyst and the efficiency of condensation of ammonia produced. The range of space velocities normally used lies between 6000–40000 V/Vh.

Being an exothermic reversible reaction improved rates of ammonia formation should result if a dropping temperature sequence is maintained within the reactor. The variation of rate with temperature at various levels of conversion for the reaction conditions outlined above is shown in Fig. 3.11. The rate data for these curves were obtained by taking the slopes of the curves for ammonia conversion versus the reciprocal of the space velocity, based on unreacted

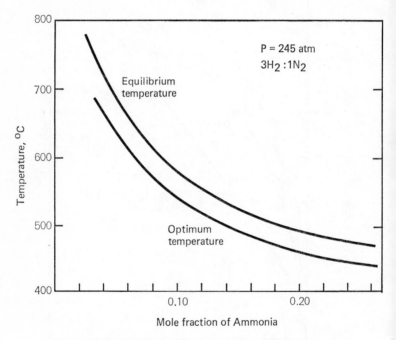

FIG. 3.13. Relation between optimum temperature and ammonia conversion (Reprinted with permission from *Chem. Eng. Science* **1,** 145 (1952))

hydrogen-nitrogen mixture, (Fig. 3.12), at each temperature and converting to the appropriate rate units. Again if an accurate Temkin equation were available these rates could be calculated analytically. The rate data are then correlated with the % NH_3 in the exit gas, obtained from Fig. 3.9 where space velocity is plotted against % NH_3 at exit as a function of temperature. Fig. 3.11 shows that the highest rates are obtained at a temperature of 500°C for ammonia percentages < 4, but above this yield a temperature of 475°C is desirable and at higher conversions lower temperatures still would be indicated.

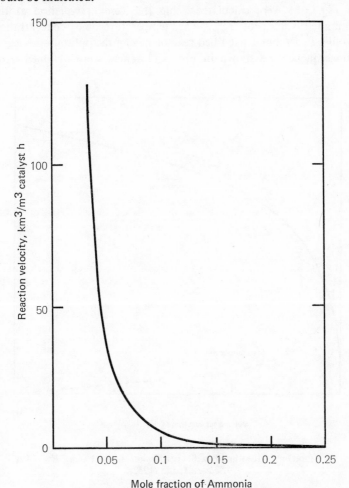

FIG. 3.14. Maximum reaction velocities at optimum conditions (Reprinted with permission from *Chem. Eng. Science* **1**, 145 (1952))

The temperature optimisation of the reaction has been considered by Annable (1952) by applying an analytical technique which utilises equations of the form of (3.3) and (3.5). For a process taking place at 245 atm. and using a 3:1, H_2:N_2 mixture, the temperature sequence shown in Fig. 3.13 was calculated using an equation similar to (3.5) and the maximum reaction velocities shown in Fig. 3.14 were calculated using the Temkin equation at the computed values of T_{max}. The temperature and concentration profiles in a tubular fixed bed reactor having the optimum temperature sequence are shown in Fig. 3.15. These were obtained from

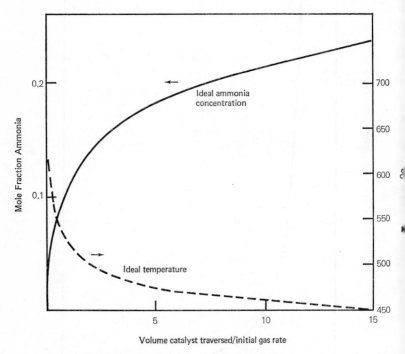

FIG. 3.15. Ideal temperature and ammonia concentration gradients (Reprinted with permission from *Chem. Eng. Science* **1,** 145 (1952))

graphical plots of $1/r_{max}$ against X, in which the volume of catalyst traversed/initial gas rate, corresponding to the given conversion X, is deduced from the area under the graph. The greater part of the conversion takes place in an initial small volume of reactor where the temperature drop is very steep.

3.3.3 OPERATING CONDITIONS FOR AMMONIA SYNTHESIS AND REACTOR TYPES

It has been established that the optimum rate of ammonia formation will take place in a reactor in which there exists a high temperature at the entrance followed by a decrease towards the exit. We have seen also that high flow rates, when conversions are kept low, result in higher space time yields being obtained than if, other things being equal, low flow rates are used. We have also seen that equilibrium yields are improved if operation is carried out at high pressures. Consideration must now be given to the question of the degree to which these desirable features can be met in practice and to what extent the design and operating conditions of reactors depend on these factors.

We have seen that there are upper limits to the space velocities which can be used, governed by the costs of circulating ammonia at high flow rates. Under the conditions of high flow rate—low conversion high degrees of recycle are required.

As well as having a favourable effect in removing the extent of reaction away from the equilibrium point increase in pressure also increases the reaction rate. However there will be an upper limit to the pressure which can be employed which will depend on the costs of compression and ease of construction of very high pressure converters. Another factor to be considered is that dissipation of heat is more of a problem at high pressures.

It has been established that operating with a falling temperature sequence is desirable and that the temperature should be as high as possible initially. Again there will be limitations to this because of the maximum temperature which the materials of construction, and particularly the catalyst, will withstand. In fact, practically, this is not so much of a problem because, as will be seen later in the section, the temperature at the entrance to the reactor cannot approach the theoretically desirable high values.

In general, it may be said that high temperature operation can be linked to high pressures and high flow rates, while lower temperatures are to be preferred at medium pressures and moderate flow rates. The efficiency and nature of the catalyst will influence the temperature at which reaction takes place; for example, very active catalysts can be employed at low temperature operation but for less active catalysts higher temperature operation will be required and,

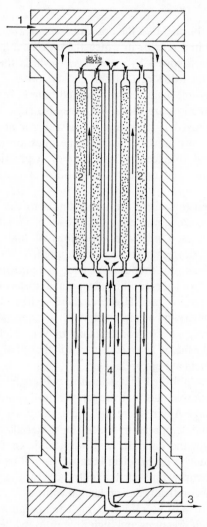

FIG. 3.16. Tube cooled ammonia synthesis converter—
countercurrent flow, 1. Feed gas entrance, 2. Catalyst beds,
3. Reacted gas exit, 4. Heat exchanger

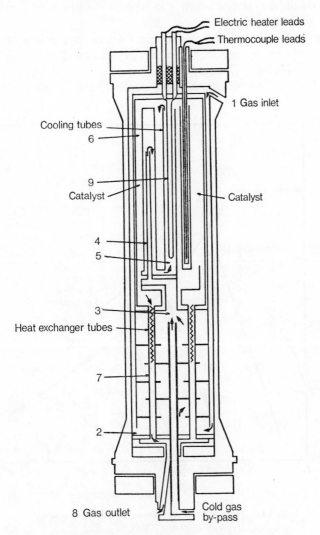

Fig. 3.17. Tube cooled ammonia synthesis converter—co-current flow, 2. Heat exchanger, 3. Mixing chamber for bypass gas, 4. Cooling tube, 5. Collecting tube, 6. Catalyst tubes, 7. Heat exchanger tubes, 9. Electric heater to control gas temperature (Reprinted with permission from *Chem. Eng. Prog.* **48**, 412 (1952))

D

as a result, the catalyst which is to be used must be more rugged in nature.

Two general types of reactor are used in ammonia synthesis: (a) tube cooled reactors and (b) interbed cooled reactors.

(a) In this type of reactor cooling tubes are inserted into the catalyst bed and the design of these cooling tubes is all important

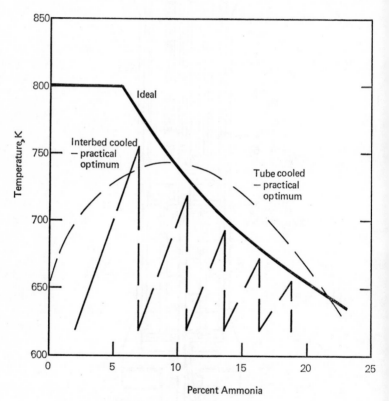

FIG. 3.18. Temperature–conversion profiles in ammonia synthesis reactors

in maintaining the desired temperatures profile within the reactor. Reactors of this type are shown in Figs. 3.16 and 3.17. In the reactor illustrated in Fig. 3.16, gases pass through the catalyst bed cooling tubes and are preheated before reversing their direction to pass down through the catalyst bed. In the other type (Fig. 3.17) a double tube

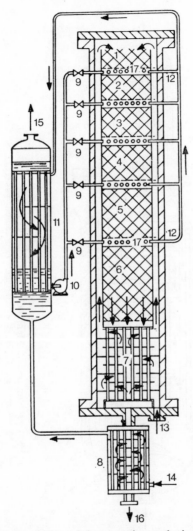

FIG. 3.19. Interbed cooled ammonia synthesis converter (Fauser–Montecatini). 1-6. Catalyst beds, 7. Heat exchanger tubes, 8. Economiser, 9. Valves to control water flowing to cooling coils, 10. Circulating pump, 11. Flash chamber, 12. Converter shell, 13. Feed gas entrance, 14. Water inlet, 15. Steam exit, 16. Reacted gas exit, 17. Cooling coils (Reprinted with permission from *Chem. Eng. Prog.* **48,** 412 (1952))

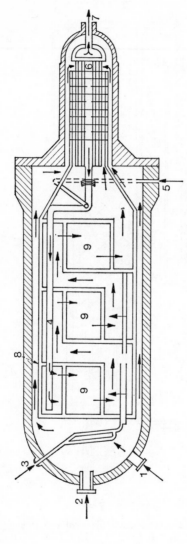

Fig. 3.20. Ammonia synthesis converter with interbed quench (Kellogg).
1. Feed gas entrance, 2. Manhole, 3. Quench gas entrance, 4. Gas transfer line,
5. Cold gas bypass, 6. Heat exchanger, 7. Reacted gas exit, 8. Cartridge,
9. Catalyst bed (Orginally published in *Hydrocarbon Processing* **51**, No. 4,
April 1972, page 127. Copyrighted by Gulf Publishing Company, Houston)

heat exchanger is employed in which gas flows down the outer part of the tube co-current to that down through the catalyst bed. Cooling tubes are sized and arranged to give optimum performance for catalyst volume, capital cost against conversion, circulating rate and power costs.

Since the temperature of the reactant gas at the entrance to the catalyst bed is determined by heat exchange considerations between unreacted and reacting gas the initial temperature will be nowhere near that required for the optimum temperature profile. In fact the temperature down the tube will rise because of the exothermic nature of the reaction. However the optimum profile is approached in the later stages of the reaction by proper design of heat transfer capacity. This is illustrated in Fig. 3.18.

(b) The second reactor type employs inter-bed cooling. In this type of reactor the synthesis reaction is allowed to proceed adiabatically within the catalyst bed but heat is then removed before the gases pass from the exit of one bed to the entrance of the next. An example of this type of reactor is the Fauser—Montecatini (Fig. 3.19).

In general the inter-bed cooled types are not so efficient in catalyst usage as are the tube-cooled reactors, requiring up to 15% more catalyst.

There is a variant of the inter-bed cooled type, sometimes called the quench converter, in which 'cold-shots' of synthesis gas are added between one catalyst bed and the next. The Kellogg reactor (Fig. 3.20) is an example of this type. In this reactor introduction of fresh gas in the inter-bed spaces decreases the ammonia concentration but the converter is optimized to give minimum catalyst volume consistent with catalyst temperature and life.

The temperature profile attained in an inter-bed cooled reactor is also shown in Fig. 3.18.

3.3.4 AMMONIA SYNTHESIS LOOP

In addition to the packed bed reactor itself there are other pieces of equipment, the design of which must be carefully carried out if efficient production is to be achieved. These constitute what is known as the synthesis loop. Two typical loops are illustrated in Fig. 3.21.

Fig. 3.21a depicts the simplest type of loop which will now be described. It is necessary to include a cooler condenser and a separator to remove liquid ammonia from the unreacted gases. Because of the relatively small conversions per pass, recirculation of gases is essential and, since the feed gases are likely to contain inert

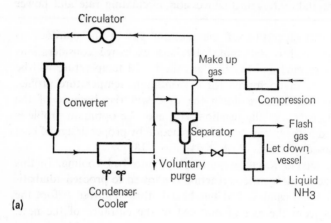

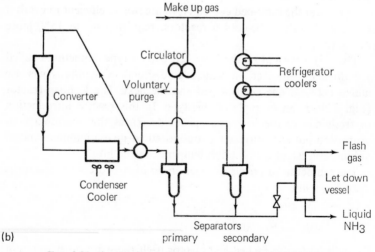

Fig. 3.21. Ammonia synthesis loops (Reprinted with permission from *Chem. Proc. Eng.*, Sept. 1965, p. 473)

materials such as methane and argon, the amounts of which will depend upon the source of the hydrogen and nitrogen, a purge may be necessary. However an involuntary purge does in fact operate since the gases will dissolve in the pressurised products and are then flashed off in reducing to atmospheric pressure. Fresh make up gas is usually added after the voluntary purge point but before the separator. One effect of the presence of inert material in the feed gas is to lower the overall theoretical thermodynamic equilibrium yield of ammonia which can be attained. This may be seen by examination of equation (3.9).

Fig. 3.21b illustrates a more complex loop type in which refrigeration of the circulating gas, followed by separation, takes place in two stages. In the first, cooler air or cooling water is used while in the second liquid ammonia is employed. The result is to raise the amount of ammonia condensed which consequently reduces the ammonia concentration in the recirculating gases from about 7 to 3% and increases the conversion/pass from 7–19% instead of 3–16%. A smaller converter can be used if the refrigeration procedure is carried out but of course the extra cost of refrigeration must be considered.

3.3.5 AMMONIA SYNTHESIS CATALYSTS AND MECHANISM OF REACTION

The basic constituent of the catalyst used for ammonia synthesis is iron. Although other metals, including Mo, Ru, W, Re, Os, could be used, iron, because of its low cost, is the universally adopted material.

The synthesis of ammonia by means of a thermal uncatalysed reaction has not been achieved; neither has it been found possible to carry out the synthesis from atomic hydrogen and molecular nitrogen. Therefore it would appear that in some way the catalyst brings the nitrogen molecule into a reactive state. The bond dissociation energy of nitrogen is 669 kJ/mol which suggests that a very high temperature would be required for dissociation. However ammonia synthesis can be carried out at temperatures as low as 500°K, which suggests an activation energy for the reaction of about 130–170 kJ/mol. The mechanism would appear to involve the chemisorption of the nitrogen molecule on to the iron surface but the chemical links which are formed are not over-strong and, after reaction, subsequent detachment of ammonia from the surface takes place readily.

The catalyst which is used in the synthesis is not, of course, pure iron. A typical formulation from which a catalyst is made contains in addition to the iron constituent (magnetite, Fe_3O_4) about 0.8% K_2O, 2% CaO, 2.5% Al_2O_3 together with smaller amounts of MgO, SiO_2 and traces of TiO_2, ZrO_2 and V_2O_5 (Bridger and Snowdon, 1970). Synthesis catalysts are made by fusing magnetite with the required small amounts of oxides (called promoters), cooling the mix, breaking it up and then screening the particles to the required size. Before use the catalyst is reduced to metallic iron by hydrogen or synthesis gas, either in the actual converter or by pre-reduction. Magnetite has a spinel structure with Fe^{2+} and Fe^{3+} in the interstices of a cubic packing of oxygen ions. During reduction the oxygen is removed but no shrinkage occurs and the resulting iron has a very porous structure suitable for use as a catalyst.

The function of the Al_2O_3 and MgO is to control the size of the iron crystallites. During fusion these oxides dissolve in the magnetite, Al^{3+} and Mg^{2+} replacing Fe^{3+} and Fe^{2+}, but during the reduction process the Al_2O_3 and MgO come out of solution and by so doing limit the further growth of the iron crystallites. The surface area of the reduced catalyst is 15–$20\ m^2/g$ as compared with $1\ m^2/g$ for the unreduced and $< 1\ m^2/g$ for pure unpromoted iron.

K_2O increases the intrinsic activity of the iron but also tends to decrease the surface area. Hence an optimum amount of K_2O must be included in the catalyst. Some of the potash reacts with the magnetite to form potassium ferrites on fusion. These are uniformly distributed in the phases between the Fe_3O_4 crystallites and on reduction potassium is distributed uniformly over the porous iron structure. Some of the K_2O reacts with excess alumina and silica during fusion to give alumino-silicates which are stable on reduction. This reduces the amount of K_2O available for ferrite formation. However the CaO and other basic promoters also react with alumina and silica to give alumino-silicates and so effectively more K_2O is free to activate the iron. CaO enhances the action of Al_2O_3 by stabilising the iron surface area and preventing sintering. Silica, although it neutralises K_2O, also has a stabilising effect like Al_2O_3.

There is obviously considerable interaction between the compounds and optimisation of the amounts to be added is necessary.

Before concluding this section it should be mentioned that the catalyst is very sensitive to poisoning by sulphur, and oxygen containing compounds such as H_2O, CO and CO_2.

REFERENCES

ADAMS, R. M. and COMINGS, E. W. 1953. *Chem. Eng. Prog.* **49**, 359.

ALLEN, P. 1965. *Chem. Proc. Eng.* **46**, 473.

ANNABLE, D. 1952. *Chem. Eng. Sci.* **1**, 145.

BRIDGER, G. W. and SNOWDON, C. B. 1970. In *Catalyst Handbook*. Wolfe Scientific Books, London, p. 126.

DYSON, D. C. and SIMON, J. M. 1968. *Ind. Eng. Chem. (Fundamentals)* **7**, 605.

GILLESPIE, L. J. and BEATTIE, J. A. 1930. *Phys. Rev.* **36**, 1008.

HARRISON, R. H. and KOBE, K. A. 1953. *Chem. Eng. Prog.* **49**, 349.

LARSON, A. T. 1924. *J. Amer. Chem. Soc.* **46**, 367.

LARSON, A. T. and DODGE, R. L. 1923. *J. Amer. Chem. Soc.* **45**, 2918.

NIELSEN, A. 1968. *An investigation on promoted iron catalysts for the synthesis of ammonia*, 3rd Ed., Jul. Gjellerups Forlag, Copenhagen.

TEMKIN, M. I., MOROZOV, N. M., and SHAPATINA, E. N. 1963. *Kinetics and Catalysis (U.R.S.S.)* **4**, 565.

TEMKIN, M. I. and PYZHEV, V. 1940. *Acta Physicochim (U.R.S.S.)* **12**, 327.

4

STEAM REFORMING OF HYDROCARBONS

4.1 Introduction to Steam Reforming

The steam reforming of hydrocarbons is a process which has assumed great importance in recent years because it can be adapted in several ways to provide different types of synthesis gas as well as fuel gas. Thus the process conditions can be modified to produce hydrogen for ammonia synthesis, hydrogen and carbon oxides for methanol synthesis and Oxo reactions and a fuel gas which contains a high proportion of methane in addition to hydrogen and carbon oxides.

The first developments in the process took place in the 1930s with the introduction of plant for the reforming of low C number saturated hydrocarbons. The process has become much more important in recent years with the introduction of improved catalysts and the wide availability of light naphtha and natural gas feedstocks.

The basis of the process is the conversion of hydrocarbons to methane, CO, CO_2 and H_2 by reaction with steam over a nickel based catalyst. A typical naphtha feed, as used in the I.C.I. process (Bridger and Wyrwas, 1967), has a boiling range of 30–180°C and contains fifty or more hydrocarbons from C_3–C_{10}, 75% of which are paraffins, 15% naphthenes, 9% aromatics and 1% unsaturates. A general observation which has been made when reforming higher saturated hydrocarbons is that, provided the contact time is long enough, the exit gas composition is that which approximately corresponds to the chemical equilibria involving the methane-steam and water gas shift reactions (equations 4.1 and 4.2):

$$CH_4 + H_2O \rightleftharpoons CO + 3H_2 \tag{4.1}$$

$$CO + H_2O \rightleftharpoons CO_2 + H_2. \tag{4.2}$$

The formation of CO_2 in the reforming reaction could be considered as an alternative to CO (equation 4.3) but only two out of the three equations are necessary to represent the overall equilibria

$$CH_4 + 2H_2O \rightleftharpoons CO_2 + 4H_2. \tag{4.3}$$

At shorter contact times significant amounts of saturated and unsaturated hydrocarbons are present. It is thought that equilibrium is quickly established with regard to the shift reaction but less quickly in the methane-steam reaction.

The composition of the final gas is determined by the temperature and pressure at the exit of the catalyst bed. These govern the equilibrium concentrations of the species present, although in instances where equilibrium is not fully reached, the concept of equilibrium approach may be employed. Here an equilibrium constant is calculated from the measured exit gas composition and the temperature corresponding to this composition computed. The difference between this temperature and the exit temperature is the equilibrium approach and is a measure of the extent of departure from the equilibrium position.

4.1.1 THERMODYNAMICS OF REFORMING

4.1.1.1 Reforming of Methane

Methane represents the simplest hydrocarbon capable of being reformed. The product compositions are calculated using a combination of two of the three equations 4.1–4.3. The assumption made is that steam/methane is high enough that no carbon is present (see 4.1.3).

The equilibrium compositions will depend on temperature, operating pressure and steam/methane ratio.

Consider that equations (4.1) and (4.2) describe the reactions taking place.

For equilibrium at a particular temperature (by applying equation 1.55)

$$K_1 = \frac{p_{CO} p_{H_2}^3}{p_{CH_4} p_{H_2O}} = \frac{n_{CO} n_{H_2}^3}{n_{CH_4} n_{H_2O}} \cdot \left(\frac{P}{\Sigma n}\right)^2 \tag{4.4}$$

where the gases are assumed to be ideal. Σn is the sum of moles of gas at equilibrium and P is the total pressure.

Also
$$K_2 = \frac{p_{CO_2} p_{H_2}}{p_{CO} p_{H_2O}} = \frac{n_{CO_2} n_{H_2}}{n_{CO} n_{H_2O}}. \tag{4.5}$$

If values of K_1 and K_2 are known at a given temperature (see

Tables 4.1 and 4.2) then for a specified total pressure and initial steam/methane ratio the number of moles of each component at equilibrium may be determined.

TABLE 4.1

Gas phase equilibrium constants
Methane-steam reaction
$CH_4 + H_2O \rightleftharpoons CO + 3H_2$

Temperature ($°K$)	$K_p = \dfrac{p_{CO}p_{H_2}^3}{p_{CH_4}p_{H_2O}}$	Temperature ($°K$)	$K_p = \dfrac{p_{CO}p_{H_2}^3}{p_{CH_4}p_{H_2O}}$
473	$4·614 \times 10^{-12}$	973	$1·214 \times 10^1$
523	$8·397 \times 10^{-10}$	998	$2·442 \times 10^1$
573	$6·378 \times 10^{-8}$	1023	$4·753 \times 10^1$
623	$2·483 \times 10^{-6}$	1048	$8·968 \times 10^1$
673	$5·732 \times 10^{-5}$	1073	$1·644 \times 10^2$
723	$8·714 \times 10^{-4}$	1098	$2·933 \times 10^2$
773	$9·442 \times 10^{-3}$	1123	$5·101 \times 10^2$
823	$7·741 \times 10^{-2}$	1148	$8·666 \times 10^2$
873	$5·029 \times 10^{-1}$	1173	$1·440 \times 10^3$
898	1.189	1198	$2·342 \times 10^3$
923	2.686	1223	$3·736 \times 10^3$
948	5·821	1248	$5·850 \times 10^3$

TABLE 4.2

Gas phase equilibrium constants
Water-gas shift reaction
$CO + H_2O \rightleftharpoons CO_2 + H_2$

Temperature ($°K$)	$K_p = \dfrac{P_{CO_2}P_{H_2}}{P_{CO}P_{H_2O}}$	Temperature ($°K$)	$K_p = \dfrac{P_{CO_2}P_{H_2}}{P_{CO}P_{H_2O}}$
473	$2·279 \times 10^2$	873	2·527
498	$1·369 \times 10^2$	898	2·199
523	$8·651 \times 10^1$	923	1·923
548	$5·714 \times 10^1$	948	1·706
573	$3·922 \times 10^1$	973	1·519
598	$2·783 \times 10^1$	998	1·361
623	$2·034 \times 10^1$	1023	1·228
648	$1·525 \times 10^1$	1048	1·113
673	$1·170 \times 10^1$	1073	1·015
698	9·165	1098	$9·295 \times 10^{-1}$
723	7·311	1123	$8·552 \times 10^{-1}$
748	5·928	1148	$7·901 \times 10^{-1}$
773	4·878	1173	$7·328 \times 10^{-1}$
798	4·069	1198	$6·822 \times 10^{-1}$
823	3·434	1223	$6·372 \times 10^{-1}$
848	2·931	1248	$5·971 \times 10^{-1}$

Suppose initially a moles of steam are present per mole of methane. If X is the conversion of methane by reaction (4.1) and Y is the conversion of CO by reaction (4.2) then at equilibrium

$$CH_4 + H_2O = CO + 3H_2$$

No. of moles $\quad 1-X \quad a-X-Y \quad X-Y \quad 3X+Y$

$$CO + H_2O = H_2 + CO_2$$

No. of moles $\quad X-Y \quad a-X-Y \quad 3X+Y \quad Y$

i.e. $\quad n_{CH_4} = 1-X \qquad n_{H_2} = 3X+Y$

$\qquad n_{H_2O} = a-X-Y \qquad n_{CO_2} = Y$

$\qquad n_{CO} = X-Y \qquad \text{Total}\,(\Sigma n) = 1+a+2X.$

Equations (4.4) and (4.5) may be written

$$K_1 = \frac{(X-Y)(3X+Y)^3}{(1-X)(a-X-Y)} \left(\frac{P}{1+a+2X}\right)^2 \tag{4.6}$$

and

$$K_2 = \frac{(Y)(3X+Y)}{(X-Y)(a-X-Y)}. \tag{4.7}$$

These equations are then solved simultaneously for X and Y. Solution is obviously complex and is best done by an iterative technique. An example on the steam reforming of methane has been given in section (1.6.2).

It is customary to report the product composition data in steam reforming reactions on a steam free basis since the steam is not a constituent in any of the synthesis gases produced or in the reformed gas when used as a fuel. The variation of reformed gas composition over a wide range of temperature, operating pressure and steam/methane ratios for methane reforming has been reported graphically (Bridger and Chinchen, 1970). This information is reproduced in Figs 4.1–4.3. The hydrogen content is obtained by difference.

It is obvious from these figures that variation in reaction conditions has a marked effect on the product composition and hence the production of different gas streams can be closely controlled by the correct choice of reaction conditions. For example, in hydrogen synthesis a very low methane content is required and it is seen that this is best achieved by operating under conditions of high temperature, low pressure and high steam/methane ratios. On the other hand, for fuel gas requirements, the product should be methane rich and this is achieved at low temperatures, high pressures and low steam/

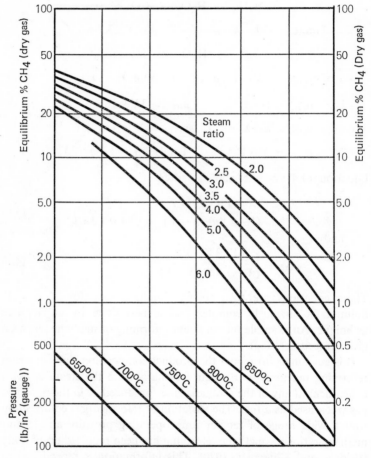

FIG. 4.1. Steam reforming of methane—equilibrium concentration of methane as a function of temperature, pressure and steam ratio (Reprinted with permission from *Catalyst Handbook*, Wolfe Scientific Books, London (1970), p. 66)

methane ratios. There is however a restriction on the lower limit of the H_2O/CH_4 ratio which can be used. This should be high enough to suppress formation of carbon on the catalyst.

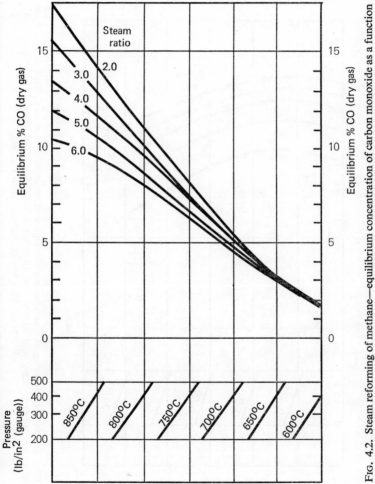

Fig. 4.2. Steam reforming of methane—equilibrium concentration of carbon monoxide as a function of temperature, pressure and steam ratio (Reprinted with permission from *Catalyst Handbook*, Wolfe Scientific Books, London (1970), p. 67).

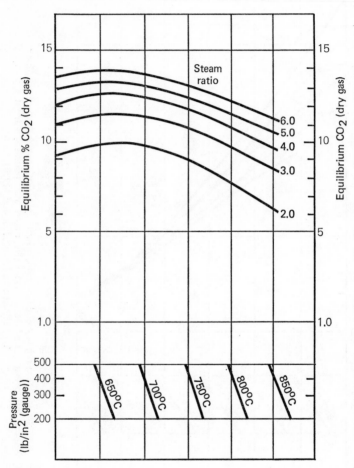

Fig. 4.3. Steam reforming of methane—equilibrium concentration of carbon dioxide as a function of temperature, pressure and steam ratio (Reprinted with permission from *Catalyst Handbook*, Wolfe Scientific Books, London (1970), p. 68).

4.1.1.2 *Reforming of Higher Hydrocarbons*

For the reforming of higher hydrocarbons, such as are present in light naphtha, it has been observed that, under typical reforming conditions, the only hydrocarbon present is methane and it is suggested that the higher hydrocarbons undergo complete reaction with water according to equations (4.8) or (4.9):

$$C_mH_{2n} + m\,H_2O \rightarrow m\,CO + (m+n)\,H_2 \qquad (4.8)$$

and

$$C_mH_{2n} + 2mH_2O \rightarrow m\,CO_2 + (2m+n)\,H_2. \qquad (4.9)$$

The steam reforming of higher hydrocarbons will then be described by either (4.8) or (4.9) together with reactions (4.1) and (4.2).

Suppose the steam/hydrocarbon ratio is a moles H_2O/mole of C_mH_{2n}. Then at equilibrium

$$C_mH_{2n} + 2mH_2O \rightarrow m\,CO_2 + (2m+n)H_2$$

No. of moles	$a-2m+X+Y$	$m-X$	$2m+n-X-3Y$

$$CO_2 + H_2 \rightleftharpoons CO + H_2O$$

No. of moles	$m-X$	$2m+n-X-3Y$	$X-Y$	$a-2m+X+Y$

$$CO + 3H_2 \rightleftharpoons CH_4 + H_2O$$

No. of moles	$X-Y$	$2m+n-X-3Y$	Y	$a-2m+X+Y$

i.e. $n_{CO} = X-Y$

$\quad\ n_{H_2O} = a-2m+X+Y$

$\quad\ n_{CO_2} = m-X$

$\quad\ n_{H_2} = 2m+n-X-3Y$

$\quad\ n_{CH_4} = Y$

Total $(\Sigma n) = a+m+n-2Y$.

Hence equations (4.4) and (4.5) may be written

$$K_1 = \frac{(X-Y)(2m+n-X-3Y)^3}{(Y)(a-2m+X+Y)} \left\{ \frac{P}{a+m+n-2Y} \right\}^2 \qquad (4.10)$$

$$K_2 = \frac{(m-X)(2m+n-X-3Y)}{(X-Y)(a-2m+X+Y)}. \qquad (4.11)$$

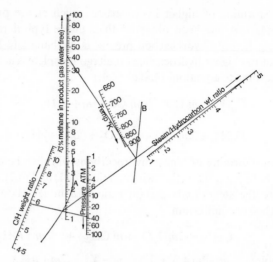

FIG. 4.4(a). Methane content in dry reformed gas

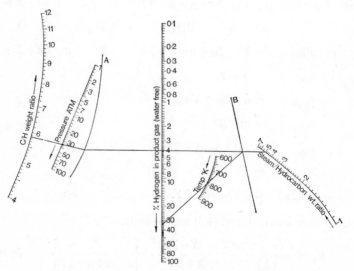

FIG. 4.4(b). Hydrogen content in dry reformed gas

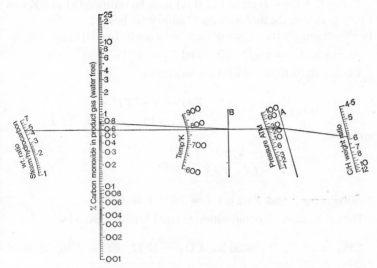

Fig. 4.4(c). Carbon monoxide content in dry reformed gas (originally published in *Hydrocarbon Processing*, **46**, No. 9, September 1967, page 1970. Copyrighted by Gulf Publishing Company, Houston)

Method of using nomographs—join C/H weight ratio to the required pressure to get a reference point on curve A. Join this point to the steam/hydrocarbon feed ratio to get another reference point on the line B. Join this point to the temperature and extend this line to the methane (a) hydrogen, (b) or carbon monoxide (c) line to read the respective composition in the water-free product gas.

Example 4.1. $n-$Hexane (C_6H_{14}) is to be reformed at 773°K and 1 atm. pressure, the moles steam/C atom ratio being 2.

For this example the values of m, n and a are 6, 7 and 12 respectively.

At 773°K, $K_1 = 9\cdot442 \times 10^{-3}$ and $K_2 = 4\cdot878$.

Equations (4.10) and (4.11) are written as

$$9\cdot442 \times 10^{-3} = \frac{(X-Y)(19-X-3Y)^3}{(Y)(X+Y)}\left(\frac{1}{25-2Y}\right)^2$$

and

$$4\cdot878 = \frac{(6-X)(19-X-3Y)}{(X-Y)(X+Y)}.$$

Solving for X and Y we get $X = 3\cdot645$ $Y = 3\cdot245$

Hence the mole % composition at equilibrium is given by

$CH_4 = 17\cdot54$, $H_2 = 30\cdot36$, $CO_2 = 12\cdot72$, $CO = 2\cdot16$, $H_2O = 37\cdot22$.

On a steam free basis the mole % composition is

$CH_4 = 27\cdot94$, $H_2 = 48\cdot36$, $CO_2 = 20\cdot26$, $CO = 3\cdot44$.

Steam free reformer gas compositions can be estimated from nomographs devised by Subramaniam (1967) which may be applied to any saturated hydrocarbon (Fig. 4.4) and graphs, similar to those of Figs 4.1−4.3 for methane, have been constructed for the equilibrium compositions of components in naphtha reforming (Bridger and Chinchen, 1970).

The importance of reaction conditions on the product composition is shown in Figs 4.5 and 4.6 for the case of a naphtha with a H/C ratio of 2·25 being reformed at 13 atm. Reaction at 1123°K with a moles steam/C atom ratio of 2·4 produces a synthesis gas containing 2·5 mole % methane, the calorific value of the gas being low since the main products are carbon monoxide and hydrogen. A low methane containing gas of this type will be suitable, after further treatment, as ammonia synthesis gas. On the other hand reforming at 733°K with a moles steam/C atom ratio of 1·6 leads to a gas containing 62·5 mole % CH_4, the gas having a calorific value of 26780 J/m^3, (measured at N.T.P.), and being suitable for towns gas.

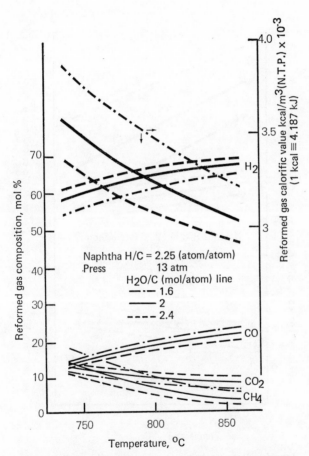

Fig. 4.5. Relation between gas composition and calorific value as a function of temperature and steam/carbon ratio when producing synthesis gas (Originally published in *Hydrocarbon Processing*, **47**, No. 2, February 1968, page 87. Copyrighted by Gulf Publishing Company, Houston)

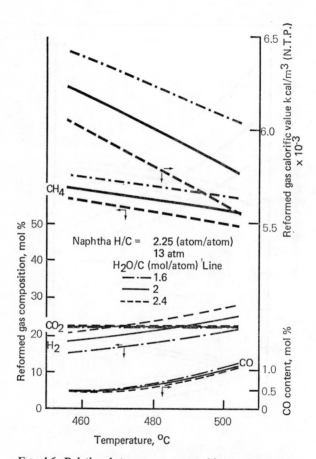

FIG. 4.6. Relation between gas composition and calorific value as a function of temperature and steam/carbon ratio when producing towns gas (Originally published in *Hydrocarbon Processing*, **47**, No. 2, February 1968, page 87. Copyrighted by Gulf Publishing Company, Houston)

4.1.2 HEAT OF REACTION IN REFORMING

An interesting consequence of the effect of reaction conditions on the position of equilibrium for reactions (4.1) and (4.2) is the resultant variation in the overall heat of reaction. The methane-steam reaction is an endothermic reaction, the heat of reaction for reactions (4.1) and (4.3) at 298°K being 205·5 kJ/mole and 165 kJ/mole respectively while the shift reaction, $CO + H_2O \rightleftharpoons CO_2 + H_2$, is exothermic, $\Delta H°$ being $-41·1$ kJ/mole at 298°K.

Reactions (4.8) and (4.9), forming CO and H_2 and CO_2 and H_2 respectively, are both endothermic and the overall heat effect will depend on the relative extent to which reactions (4.1) and (4.2) proceed, since (4.8) and (4.9) are assumed to go to completion. At high temperatures the equilibrium position of equation (4.1) lies to the right while that of (4.2) is to the left. The net result is an overall endothermic reaction which therefore necessitates the addition of heat to the reforming reactor at high temperatures. On the other hand, at temperatures where the formation of fuel gas is favoured a smaller amount of methane is reformed resulting in a reduced need for heat addition. At such temperatures the overall process can be slightly exothermic.

These statements can be illustrated by two cases reported in the literature (Bridger and Chinchen, 1970). In the first at 1073°K and 20 atm. a value of 100·4 kJ is given for the heat change in the steam reforming of naphtha (moles steam/C atom = 3/1) according to the following stoichiometric equation

$$CH_{2\cdot2} + 3H_2O \rightleftharpoons 0·2\,CH_4 + 0·4CO + 0·4CO_2 + 1·94H_2 + 1·81H_2O,$$

i.e. the reaction is substantially endothermic. On the other hand for the reaction taking place at 723°K and 30 atm. the heat of reaction is $-47·6$ kJ (moles steam/C atom = 2/1) for the stoichiometry written as

$$CH_{2\cdot2} + 2H_2O = 0·75\,CH_4 + 0·25\,CO_2 + 0·14\,H_2 + 1·5\,H_2O.$$

The reaction is most endothermic when the whole of the hydrocarbon is reformed to carbon oxides and hydrogen and becomes less so, and eventually exothermic, as the amount of methane present increases.

4.1.3 FORMATION OF CARBON

For effective operation of the catalyst it is essential that carbon deposition does not take place. Reforming of higher hydrocarbons in particular is prone to result in carbon formation, presumably due to increased ease of pyrolysis.

Carbon could also be formed according to the Boudouard reaction

$$2CO \rightleftharpoons CO_2 + C \qquad (4.12)$$

or by the reaction

$$CO + H_2 \rightleftharpoons C + H_2O. \qquad (4.13)$$

The necessary conditions for the suppression of carbon formation at equilibrium are that

$$\frac{p_{CO_2}}{(p_{CO})^2} > K_{12}, \qquad \frac{p_{H_2O}}{p_{CO} p_{H_2}} > K_{13}$$

and

$$\frac{(p_{H_2})^n}{p_{C_m H_{2n}}} > K_{14} \qquad \text{where } K_{14} \text{ is the equilibrium}$$

constant for

$$C_m H_{2n} \rightleftharpoons mC + n\, H_2. \qquad (4.14)$$

Where the hydrocarbon being reformed is methane, $(p_{H_2})^2/p_{CH_4}$ must be $> K_{14}$.

These conditions will be satisfied by the choice of a high enough steam/hydrocarbon ratio.

In the example given previously, for the reforming of $C_6 H_{14}$ at $773°K$,

$$\frac{(p_{H_2})^2}{p_{CH_4}} = \frac{(30.36)^2}{17.54 \times 100} = 0.528 \qquad (K_{14} = 0.12)$$

$$\frac{p_{H_2O}}{p_{CO}; p_{H_2}} = \frac{37.22 \times 100}{2.16 \times 30.36} = 56.76 \qquad (K_{13} = 13)$$

and

$$\frac{p_{CO_2}}{(p_{CO})^2} = \frac{12.72}{(2.16)^2} \times 100 = 273 \qquad (K_{12} = 68).$$

Hence in accordance with the assumption no carbon will be deposited.

The minimum steam ratio will be obtained when the equilibrium equalities are satisfied. The limiting steam ratio could be calculated directly from equilibrium expressions for reactions (4.1) and (4.12) since these embrace all components including carbon.

It should be emphasised that the above calculation assumes that the system is at equilibrium. It is important to be sure that at no point during the course of the reaction, before equilibrium is attained,

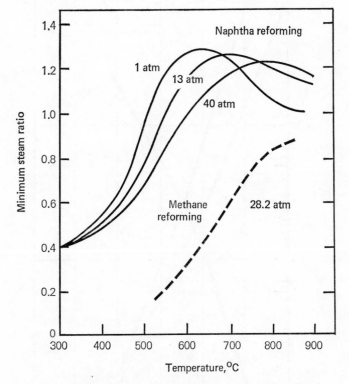

FIG. 4.7. Thermodynamic minimum steam ratios (Reprinted with permission from *Catalyst Handbook*, Wolfe Scientific Books, London (1970), p. 74

should the above ratios fall below the respective equilibrium constant values. Whether this condition is achieved can be ascertained only by a full analysis of the kinetics of the various reactions taking place.

Some values for the minimum steam-hydrocarbon ratios under various conditions were reported by Dent (1946) together with equilibrium constants for the reactions involved (Figs 4.7 and 4.8).

It has been reported that addition of alkali metal salts to the catalyst allows a much closer approach to the thermodynamic minimum steam ratio since these enhance the rate of carbon removal in reactions (4.12) and (4.13).

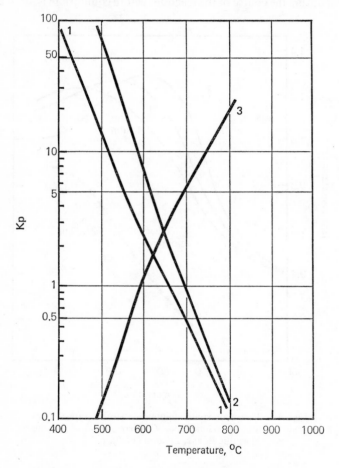

FIG. 4.8. Equilibrium constants for carbon forming reactions

(1) $CO + H_2 \rightleftharpoons C + H_2O$
(2) $2CO \rightleftharpoons C + CO_2$
(3) $CH_4 \rightleftharpoons C + 2H_2$

(Reprinted with permission from *Catalyst Handbook*, Wolfe Scientific Books, London (1970), p. 74

4.1.4 KINETICS OF STEAM REFORMING REACTIONS

There is still considerable disagreement regarding the mechanism of steam reforming and the form of the rate equations to describe the kinetics of the process. This may in part be due to variations in the types of catalyst used in different investigations but also because of probable differences in mechanism at high and low temperatures.

The initial products in the steam reforming of methane have been postulated by Akers *et al.* (1955) to be carbon monoxide, carbon dioxide and hydrogen formed according to

$$CH_4 + H_2O \rightleftharpoons CO + 3H_2 \tag{4.1}$$

and

$$CH_4 + 2H_2O \rightleftharpoons CO_2 + 4H_2. \tag{4.3}$$

The shift reaction is then considered to follow as a slow secondary process.

On the other hand Bodrov *et al.* (1964) postulated that the primary products were carbon monoxide and hydrogen, carbon dioxide resulting from oxidation of the monoxide.

Yet again Moe and Gerhard (1965) suggested that the important reactions were

$$CH_4 + 2H_2O \rightleftharpoons CO_2 + 4H_2 \tag{4.3}$$

and

$$CO_2 + CH_4 \rightleftharpoons 2CO + 2H_2. \tag{4.15}$$

Fortunately if the reactions proceed until an equilibrium position is reached a knowledge of the sequence of reaction steps is not so crucial as if the process were determined by kinetic considerations.

With regard to a kinetic equation for the process Akers and Camp (1955), using a nickel on kieselguhr catalyst at 911°K and 1 atm. pressure, found the reaction to be first order with respect to methane and zero order in steam. Moe and Gerhard (1965) represented the rate equation by

$$r = k \left(K_3 \, p_{CH_4} \, p_{H_2O}{}^2 - p_{H_2}{}^4 p_{CO_2} \right)$$

where K_3 is the equilibrium constant for reaction (4.3). Bodrov *et al.* (1964), for reaction at 1073–1173°K and 1 atm. pressure on a nickel foil catalyst chosen to reduce diffusion effects, postulated a rate equation of the form

$$r = \frac{k\,p_{CH_4}}{1+a(p_{H_2O}/p_{H_2})} + b\,p_{CO}.$$

However on repeating the experiments with nickel deposited on
α-alumina, in agreement with Akers, they found that the simple
equation $r = kp_{CH_4}$ satisfactorily represented the results.

Likewise for the reforming of higher hydrocarbons there is a fair
measure of disagreement regarding the mechanism and kinetics.

Bhatta and Dixon (1967) studied the steam reforming of n-butane
at a low temperature of 698–748°K and 30 atm. and found that the
reaction products of methane, hydrogen, carbon monoxide, carbon
dioxide and water were at equilibrium at all conversions of the
butane. The reaction was first order with respect to water but zero
in butane suggesting that the butane initially covers the surface of
the catalyst.

Phillips *et al.* (1969) have studied the steam reforming of n-
hexane at 623–773°K and 14 atm. on a co-precipitated nickel-
alumina catalyst of a different type to that employed by Bhatta.
The reaction was found to change from zero order in hydrocarbon
at the low temperatures to 0·3 at higher and this was interpreted in
terms of a Langmuir kinetic mechanism where the apparent order
of reaction increases from $0 \rightarrow 1$ as the surface coverage decreases.
The reaction was zero order in steam. No lower paraffin products
were detected in the decomposition of n-hydrocarbons but some
when branched chain compounds were reacted. They postulated
that the hydrocarbon was adsorbed by a Langmuir mechanism and a
normal paraffin reacts on the surface to give $(CH)_x$ which then
reacts with steam, this being the rate controlling step. The steam is
strongly adsorbed and the products are hydrogen and carbon oxides
which desorb and interact to give equilibrium gas mixtures. For the
branched chain hydrocarbons the rate controlling step is the formation of $(CH)_x$. The work of Bhatta was interpreted as involving
strong adsorption of butane in agreement with their own results
at low temperatures. This is in contrast to the weaker adsorption
of the hydrocarbon in methane reforming which gives rise to a first
order dependence as noted by Akers and Bodrov.

It may be that at the low temperatures (<873°K) the reaction
involves catalytic cracking of hydrocarbons producing methane and
unsaturated compounds. Unsaturated compounds have been detected

at very low contact times. These then react with steam to give hydrogen and oxides of carbon as suggested by Phillips. At higher temperatures a large part of the cracking may be thermal. Evidence for this exists because there is an increase in the measured energy of activation as the temperature increases. Following reaction with steam and desorption, equilibration between H_2, CO, CO_2, H_2O and CH_4 takes place.

This section illustrates well the many uncertain features about the kinetics of most catalytic reactions. For design purposes it is usually the case that empirical rate equations have to be postulated and these should be used only with the catalyst being employed and over the range of reaction conditions in which experimental measurements have been made.

4.2 Further Treatment of Reformed Gases for Synthesis Gas Preparation

The importance of variation of temperature, pressure and hydrocarbon/steam ratio on the reformed exit gas composition has been demonstrated. For example, at high temperatures the hydrogen content is high and the methane content low. If it is desired to produce a hydrogen feed gas, further treatment must be carried out to remove all the methane and carbon oxides; on the other hand, if a methanol synthesis gas is required it is necessary that the carbon oxide—hydrogen ratio should be of the required stoichiometric order and carbon in some form may have to be added to achieve this ratio.

In general therefore further treatment or modification of the reformed gas stream must be made to provide the required synthesis gas. The production of hydrogen for ammonia synthesis from reformed gas streams will now be considered.

4.2.1 HYDROGEN FOR AMMONIA SYNTHESIS

4.2.1.1 Secondary Reforming

It is essential in ammonia synthesis that the methane content of the feed gas be reduced as low as possible. This is achieved in a secondary reformer which operates at a temperature 150–200°C in excess of that in the primary reformer. At a temperature of the order of 1173–1373°K the equilibrium dry gas methane content is reduced to

$<0.5\%$ as can be ascertained from an extrapolation of Fig. 4.1. This high temperature can be achieved by utilising the heat released in the exothermic reaction involving the combustion of some of the methane contained in the gas from the primary reformer. The extent of the combustion reaction depends on the oxygen content, which in turn depends upon the requirement that the amount of nitrogen introduced in the air will be such that the gas entering the ammonia synthesis reactor will contain hydrogen to nitrogen in the ratio of 3:1. If the secondary reformer is operated adiabatically the entrance temperature must be chosen consistent with the required rise in temperature appropriate to the desired outlet methane concentration, and to the demand that the $H_2:N_2$ shall be 3:1. Hence the important factors to be controlled in secondary reformer operation are operating pressure, steam to carbon ratio, air to carbon ratio and inlet and outlet temperatures.

A typical mole $\%$ gas composition leaving the primary reformer would be H_2 34·3; CO 6·4; CO_2 8·3; CH_4 5·0; H_2O 45·8; N_2 0·2, while that leaving the secondary reformer would be H_2 31·5; CO 8·6; CO_2 6·5; CH_4 0·2; H_2O 40·3; N_2 12·7; A 0·2.

4.2.1.2 Water-Gas Shift Reaction

The next stage in the preparation of hydrogen synthesis gas is to remove the carbon oxides. The concentration of carbon monoxide present in the secondary reformer exit stream can be reduced by operating the water-gas shift reaction. This has the effect of converting the monoxide to the dioxide and increasing the hydrogen content at the same time. The carbon dioxide in the gas stream can be subsequently removed by absorption in an appropriate solvent.

The water gas shift reaction has been introduced in section 4.1

$$CO + H_2O \rightleftharpoons CO_2 + H_2. \qquad (4.2)$$

It is reversible and exothermic to the extent of 41·1 kJ/mol at 298°K. Hence the equilibrium concentration of CO at the exit of the reactor will fall as the temperature decreases but, as usual, the reaction rate will be greater the higher the temperature, and so some compromise has to be reached regarding the temperature of operation of the shift reaction.

Values of the equilibrium constant, as a function of temperature, have been given in Table 4.2. Because there is no change in the number of moles during reaction the extent of conversion is independent of the operating pressure but will depend upon the carbon monoxide/steam ratio.

Suppose that the shift reactor feed contains a moles hydrogen, b moles carbon monoxide and c moles carbon dioxide. If X is the degree of conversion and d the moles of steam added, the number of moles of each component at equilibrium will be given by

$$CO \; + \; H_2O \rightleftharpoons CO_2 \; + \; H_2 \qquad (4.2)$$

No. of moles $b-bX$ $d-bX$ $c+bX$ $a+bX$

Hence
$$K_2 = \frac{(c+bX)(a+bX)}{(b-bX)(d-bX)}$$

and
$$d = \frac{(c+bX)(a+bX)}{(b-bX)K_2} + bX. \qquad (4.16)$$

Example 4.2. Suppose that the gas entering the shift converter contains 34·9 moles H_2, 7·2 moles CO and 5·4 moles CO_2 and 90% conversion of the monoxide is being sought. At 673°K how much steam is required? (At 673°K, $K_2 = 11\cdot7$).

Solution. By equation (4.16)

$$d = \frac{(5\cdot4 + 0\cdot9 \times 7\cdot2)(34\cdot9 + 0\cdot9 \times 7\cdot2)}{7\cdot2(1 - 0\cdot9)11\cdot7} + (7\cdot2 \times 0\cdot9)$$

$$= 58\cdot36 + 6\cdot48 = 64\cdot84,$$

i.e. 64·84/7·2 = 9·01 moles H_2O/mole of CO initially present.

The final composition of the converted gas is

$$CO = 0\cdot72 \text{ moles}; \; H_2 = 41\cdot38 \text{ moles}$$

$$CO_2 = 11\cdot88 \text{ moles}; \; H_2O = 58\cdot36 \text{ moles}.$$

Fig. 4.9 shows the relationship between % CO in the converted gas and moles steam/mole of CO initially as a function of temperature for the gas composition given in Example 4.2.

It can be seen from this graph that, as the degree of conversion approaches 100%, the steam requirement becomes very large. This large steam requirement makes single-stage removal of carbon monoxide an impracticable process. In addition, because of the exothermic nature of the reaction, the temperature rises markedly in the reactor if operated adiabatically and this results in a final equilibrium concentration of monoxide which is higher than desirable. For these reasons the shift reactor invariably incorporates more than one stage.

The process is likely to take place over an iron catalyst, promoted with chromium and occasionally other metal oxides, at a temperature in the range 620–770°K and up to 30 atm. pressure. In a typical

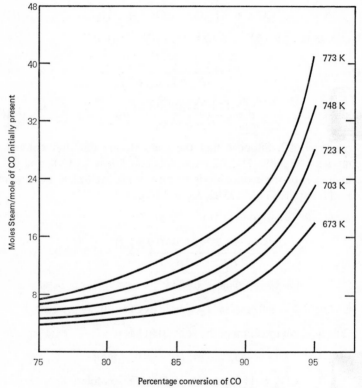

Fig. 4.9. Steam requirement in shift conversion

conversion the gas might enter the first bed of a two stage adiabatic converter at about 640°K and leave at 720°K, the carbon monoxide content being 4–5%. The exit gas is cooled in a heat recovery unit or quenched by injection of condensate before passing into the second stage of the converter. The degree of conversion in the second stage is much less than in the first and hence the temperature rise is much smaller. Consequently a lower carbon monoxide equilibrium concentration can be achieved. The exit stream from the second bed might contain 2–2·5% CO and the temperature would be about 670°K.

In older forms of operation the carbon dioxide is then removed by absorption before passing to a second shift converter which is likely to operate at a lower temperature than the first. The gas entering this converter will contain about 1% carbon dioxide and is saturated with water vapour to aid in reducing the monoxide content to <0·5%.

An alternative process, introduced more recently, employs a low temperature shift converter after the first high temperature converter (Campbell *et al.*, 1970). This process has become feasible through the introduction of a Cu-Zn catalyst on a refractory base which operates at temperatures as low as 480–510°K. With this type of catalyst there is no necessity for inter-stage carbon dioxide removal and the low temperature employed results in the attainment of low equilibrium carbon monoxide concentrations and no great need for the use of a saturator/cooler system. A typical percentage gas composition from the exit of a shift converter processing ammonia synthesis gas would be H_2 39·6; CO 0·3; CO_2 14·8; CH_4 0·3; H_2O 32·0; N_2 12·7; A 0·3.

Example 4.3 *Multistage Shift Converter.* (a) Reformer gas having a steam free mole % composition of $H_2 = 52·8$, CO = 14·4, $CO_2 = 10·9$, CH_4 + inerts = 21·9, is fed to the first stage of a two stage converter in which 60% conversion of the monoxide takes place. The temperature at the exit of this bed is 723°K. Following quenching by condensate the gas enters the second stage where 50% conversion of the remaining monoxide takes place and the gas leaves the first converter at 673°K. (At 723°K, $K_2 = 7·311$ and at 673°K, $K_2 = 11·7$).

The moles of steam/mole of CO required in each stage is to be calculated and the steam free exit composition from the converter to be determined.

Solution. It is assumed that equilibrium is approached in each stage although in actual practice this might not be fully attained.

The steam addition required is calculated by applying equation (4.16).

First bed. At 723°K, $K_2 = 7·311$.

E

Moles steam required

$$d = \frac{(10 \cdot 9 + 14 \cdot 4 \times 0 \cdot 6)(52 \cdot 8 + 14 \cdot 4 \times 0 \cdot 6)}{14 \cdot 4(1 - 0 \cdot 6)7 \cdot 311} + (14 \cdot 4 \times 0 \cdot 6)$$
$$= 37 \cdot 15,$$

or $37 \cdot 15/14 \cdot 4 = 2 \cdot 58$ moles H_2O/mole of initial CO.

The mole percentage exit composition is

$$CO = 4 \cdot 2 \quad CO_2 = 14 \cdot 25 \quad H_2 = 44 \cdot 8 \quad H_2O = 20 \cdot 78$$
$$CH_4 + \text{inerts} = 15 \cdot 97.$$

Second bed. At $673°K$, $K_2 = 11 \cdot 7$.

On a steam free basis the gas composition entering the second bed is

$$CO = 5 \cdot 3 \quad CO_2 = 17 \cdot 99 \quad H_2 = 56 \cdot 55 \quad \text{others} = 20 \cdot 16.$$

Moles steam required

$$d = \frac{(17 \cdot 99 + 5 \cdot 3 \times 0 \cdot 5)(56 \cdot 55 + 5 \cdot 3 \times 0 \cdot 5)}{5 \cdot 3(1 - 0 \cdot 5)11 \cdot 7} + (5 \cdot 3 \times 0 \cdot 5)$$
$$= 42 \cdot 06,$$

or $42 \cdot 06/5 \cdot 3 = 7 \cdot 93$ moles steam/mole of initial CO.

The mole percentage exit composition is

$$CO = 1 \cdot 87 \quad CO_2 = 14 \cdot 53 \quad H_2 = 41 \cdot 67 \quad H_2O = 27 \cdot 74$$
$$\text{others} = 14 \cdot 19.$$

(b) The carbon monoxide is then further reduced either by

(i) removal of the carbon dioxide formed and conversion in a second reactor or

(ii) by conversion in a low temperature shift reactor.

(i) Suppose 99 % of the CO_2 is scrubbed out and the exit gases leave the second converter at $673°K$.

The mole % composition of gases entering the converter on a steam free basis will be

$$CO = \frac{1 \cdot 87}{57 \cdot 875} \times 100 \quad CO_2 = \frac{0 \cdot 145}{57 \cdot 875} \times 100 \quad H_2 = \frac{41 \cdot 67}{57 \cdot 875} \times 100$$
$$= 3 \cdot 23 \qquad\qquad = 0 \cdot 25 \qquad\qquad = 72 \cdot 00$$

$$others = \frac{14 \cdot 19}{57 \cdot 875} \times 100$$

$$= 24 \cdot 52.$$

For 80% conversion, moles steam required

$$d = \frac{(0 \cdot 25 + 3 \cdot 23 \times 0 \cdot 8)(72 + 3 \cdot 23 \times 0 \cdot 8)}{3 \cdot 23(1 - 0 \cdot 8) \times 11 \cdot 7} + 3 \cdot 23 \times 0 \cdot 8$$

$$= 30 \cdot 55,$$

or $30 \cdot 55/3 \cdot 23 = 9 \cdot 46$ moles steam/mole CO.

The mole % exit composition is

$$CO = 0 \cdot 50 \quad CO_2 = 2 \cdot 17 \quad H_2 = 57 \cdot 13 \quad H_2O = 21 \cdot 42$$

$$others = 18 \cdot 78.$$

(ii) Low temperature shift reactor.

Suppose the exit temperature is 543°K at which $K = 61 \cdot 9$.

The mole % composition of gases entering converter on a steam free basis is

$$CO = 2.58 \quad CO_2 = 20 \cdot 11 \quad H_2 = 57 \cdot 67 \quad others = 19 \cdot 64.$$

For CO at exit to be about 0·5 mole % assume $X = 0 \cdot 75$, and

$$d = \frac{(20 \cdot 11 + 2 \cdot 58 \times 0 \cdot 75)(57 \cdot 67 + 2 \cdot 58 \times 0 \cdot 75)}{2 \cdot 58(1 - 0 \cdot 75)61 \cdot 9} + 2 \cdot 58 \times 0 \cdot 75$$

$$= 34 \cdot 85,$$

or $34 \cdot 85/2 \cdot 58 = 13 \cdot 51$ moles steam/mole CO.

The mole % exit composition is

$$CO = 0 \cdot 48 \quad CO_2 = 16 \cdot 345 \quad H_2 = 44 \cdot 20 \quad H_2O = 24 \cdot 41$$

$$others = 14 \cdot 565.$$

REFERENCES

AKERS, W. W. and CAMP, D. P. 1955. *A.I.Ch.E.J.* **1**, 471.

BHATTA, K. S. M. and DIXON, G. M. 1967. *Trans. Farad. Soc.* **63**, 2217.

BODROV, N. M., APEL'BAUM, L. O., and TEMKIN, M. I. 1964. *Kinetics and catalysis* (*U.R.S.S.*) **5**, 614.

BRIDGER, G. W. and CHINCHEN, G. C. 1970. In *Catalyst Handbook*. Wolfe Scientific Books, London, p. 64.

BRIDGER, G. W. and WYRWAS, W. 1967. *Chem. Proc. Eng.* **48**, 101.

CAMPBELL, J. S., CRAVEN, P., and YOUNG, P. W. 1970. In *Catalyst Handbook*. Wolfe Scientific Books, London, p. 97.

DENT, F. J. 1946. *49th Report of Joint Res. Committee, Trans. Inst. Gas. Eng.* (1945–46).

MOE, J. M. and GERHARD, E. R. 1965. Preprint 36d, 56th Natl. Meeting, A.I.Ch.E., May 16–19, 1965.

PHILLIPS, T. R., MULHALL, J., and TURNER, G. E. 1969. *J. Catalysis* **15**, 233.

SUBRAMANIAM, T. K. 1967. *Hydro. Proc.* **46**, (9), 169.

5

HYDROCARBON CRACKING AND OLEFINE MANUFACTURE

5.1 Introduction

The olefinic hydrocarbons are of supreme importance as raw materials in the organic chemical industry. Ethylene and propylene, which are consumed in the largest tonnage amounts, are used as starting materials for the synthesis of a wide variety of products including polyethylene, ethylene glycol, polypropylene, phenol via the cumene route and acrylonitrile. The successful utilisation of these simple olefines in synthesis is only possible because they can now be produced cheaply and in large amounts through the controlled pyrolysis of readily available saturated hydrocarbons.

In the United States, ethane and propane, present in natural gas, have been largely used as feedstocks for the pyrolysis process but in Western Europe, where this type of gas is not indigenous, cracking of light naphtha, a refinery distillate with a high paraffinic content, is more generally used. It is undoubtedly a simpler task to specify the process conditions for the pyrolysis of single compounds such as ethane or propane than for a complex material such as naphtha, and the analysis given in this chapter will deal more especially with the cracking of these two compounds. In hydrocarbon cracking, ethylene has in the past been the desired compound and other unsaturates have been produced in amounts which have exceeded demand, but processes based on propylene are increasing markedly in importance.

5.2 Thermodynamics of the Cracking Process

Because of the variety of products encountered in the cracking of even the simplest hydrocarbons, the cracking process is unquestionably complex. However, based on the observed amounts of the most abundant products of reaction, the following important primary reactions may be written down to describe the decomposition of ethane, propane and n-butane:

Ethane   Propane

$C_2H_6 \rightleftharpoons C_2H_4 + H_2$ $C_3H_8 \rightleftharpoons C_3H_6 + H_2$

$C_3H_8 \rightleftharpoons C_2H_4 + CH_4$

n-Butane

$C_4H_{10} \rightleftharpoons C_4H_8 + H_2$

$C_4H_{10} \rightleftharpoons C_3H_6 + CH_4$

$C_4H_{10} \rightleftharpoons C_2H_4 + C_2H_6.$

Table 5.1 lists the standard heat and free energy changes for the gas reactions of ethane and propane at 298°K. It is seen that these reactions are strongly endothermic and the free energy changes are highly unfavourable for product formation at this low temperature. Values of the standard free energy changes at higher temperatures are also shown in Table 5.1. As discussed in Chapter 1 the temperature at which the change becomes negative may be used as an indication of the approximate process operating temperature. For example, for ethane decomposition $\Delta G°$ attains a value of -4.686 kJ/mol at 1100°K but the standard free energy changes become negative at lower temperatures for the cracking of higher hydrocarbons, indicating that cracking would be successful at lower process temperatures than in the case of ethane.

In addition to the primary cracking reactions many secondary reactions involving products may be postulated; for example

$C_2H_6 + H_2 \rightleftharpoons 2CH_4$ $C_3H_8 + H_2 \rightleftharpoons C_2H_6 + CH_4$

$C_2H_4 + C_2H_6 \rightleftharpoons C_3H_6 + CH_4$ $C_3H_6 + CH_4 \rightleftharpoons C_4H_{10}.$

The standard free energy changes for these reactions are also listed in Table 5.1 and it is seen that, generally speaking, product formation is highly favoured at the temperatures at which the primary reactions proceed. This finding is of great significance since these reactions in many instances reduce the yield of the desired olefines.

It should be emphasised that the reactions postulated here are not meant to indicate any specific mechanism but account only for the overall observed reaction products.

TABLE 5.1

Standard heat and free energy changes for gas reactions of ethane and propane at 298°K

Reaction	ΔH°(kJ/mol)				ΔG°(kJ/mol)			
	298°K	800°K	1000°K	1200°K	298°K	800°K	1000°K	1200°K
$C_2H_6 \rightleftharpoons C_2H_4 + H_2$	136·95	143·46	144·28	144·43	101·01	35·86	8·87	−18·24
$C_3H_8 \rightleftharpoons C_3H_6 + H_2$	124·26	129·20	129·47	129·03	86·20	18·26	−9·46	−37·07
$C_3H_8 \rightleftharpoons C_2H_4 + CH_4$	81·28	79·73	78·14	76·36	40·82	−27·24	−53·64	−79·79
$C_2H_6 + H_2 \rightleftharpoons 2CH_4$	−65·03		−73·55		−68·7	−71·2	−70·75	−70·08
$C_2H_4 + C_2H_6 \rightleftharpoons C_3H_6 + CH_4$	−22·05		−22·28		−23·3	−25·73	−26·52	−27·36
$C_3H_8 + H_2 \rightleftharpoons C_2H_6 + CH_4$	−55·67		−66·15		−60·2	−63·09	−62·51	−61·55
$C_3H_6 + CH_4 \rightleftharpoons n\text{-}C_4H_{10}$	−71·71		−66·36		−29·1	41·63	68·87	95·56

From equilibrium thermodynamics it may also be deduced that in the primary hydrocarbon cracking reactions, equilibrium product concentrations are increased by a reduction in pressure or by the addition of inert material, since reaction leads to an increase in number of product molecules over reactants. For this reason steam is added to the feed material, and in fact steam addition is found to be beneficial in a number of other ways which will be discussed later. Cracking reactions in general are carried out at pressures just in excess of atmospheric.

5.3   Kinetics of Hydrocarbon Cracking Reactions

The kinetics of high temperature pyrolysis of simple saturated hydrocarbons have been entensively studied for many years for two reasons; firstly, to provide suitable rate equations and rate data for reactor design purposes and, secondly, because these chemical molecules are suitable models for the study of the mechanism of decomposition of chemical species. For the latter purpose it has been recognised that the degree of decomposition must be kept small to reduce the importance of secondary reactions and ease mechanistic interpretation, a requirement which is clearly at variance with the demands of the first type of study. Despite this divergence of emphasis there is no doubt that physical chemists, engaged on the second approach, have provided valuable information in creating an understanding of the overall process. The first part of this section will highlight important features of the kinetics of the pyrolyses as deduced from mechanistic studies, while the second will consider the more empirical observations made at higher degrees of conversion.

(a) Thermal cracking reactions are chain reactions involving chain initiation, propagation and termination steps. Because of the lower bond dissociation energies, chain initiation through C—C bond splitting far outweighs that involving C—H bond rupture. Thus the following initiation steps predominate in the case of the three simple hydrocarbons: ethane, propane and n-butane

$$C_2H_6 \rightarrow 2CH_3 \cdot \tag{5.1}$$

$$C_3H_8 \rightarrow CH_3 \cdot + C_2H_5 \cdot \tag{5.2}$$

$$nC_4H_{10} \rightarrow 2C_2H_5 \cdot \tag{5.3a}$$

$$\rightarrow CH_3 \cdot + C_3H_7 \cdot \tag{5.3b}$$

The type of reactions involved in the chain propagating steps were first put forward in 1934 by Rice and Herzfeld (1934) and these are still recognised as valid today (Leathard and Purnell, 1970). The two important steps may be written, in general form, as

$$R_1 \cdot \rightarrow R_2 \cdot + \text{olefine} \qquad (5.4)$$

$$R_3 \cdot + RH \rightarrow R_3 H + R_4 \cdot \qquad (5.5)$$

where $R_1 \cdot$ and $R_4 \cdot$ are 'large' radicals, which can have several isomers, and $R_2 \cdot$ and $R_3 \cdot$ are 'small' radicals. In the decomposition of $R_1 \cdot$, C—C bond splitting is favoured over C—H splitting; only when C—C bond breaking cannot readily take place will H atoms be formed as, for example, with ethyl and 2-propyl radicals where H atoms and ethylene and propylene are formed respectively. For the cases of ethane and propane the following chain propagating reactions may be postulated:

Ethane

$$C_2 H_5 \cdot \rightarrow H \cdot + C_2 H_4 \qquad (5.6)$$

$$H \cdot + C_2 H_6 \rightarrow H_2 + C_2 H_5. \qquad (5.7)$$

Propane

$$CH_3 CH_2 CH_2 \cdot \rightarrow CH_3 \cdot + C_2 H_4 \qquad (5.8)$$

$$CH_3 \cdot + C_3 H_8 \rightarrow CH_4 + C_3 H_7 \cdot \qquad (5.9)$$

$$CH_3 CH \cdot CH_3 \rightarrow H \cdot + C_3 H_6 \qquad (5.10)$$

$$H \cdot + C_3 H_8 \rightarrow H_2 + C_3 H_7 \cdot. \qquad (5.11)$$

In the majority of cases, reactions of type (5.5) involving H· and $CH_3 \cdot$ radicals are slower than type (5.4). As a consequence small radicals will predominate resulting in these being the important species in chain termination, for example, $CH_3 \cdot$ radicals in propane. However, an exception to this generalisation is the ethane case where reaction (5.7) is faster than the decomposition reaction (5.6). For ethane, ethyl radicals will be the chain ending species.

For propane, reactions (5.10) and (5.11) have been included in addition to the expected reactions (5.8) and (5.9). In certain circumstances (5.10) may be the rate limiting step because the decomposition of the 2-propyl radical involves C—H bond breaking. In this event the 2-propyl radical will take part in the chain ending step.

The classical Rice-Herzfeld mechanism leading to 3/2 order

kinetics, as experienced in the case of propane, will now be considered. The participation of the 2-propyl radical is not included. The mechanism may be written:

$$C_3H_8 \xrightarrow{k_1} CH_3\cdot + C_2H_5\cdot \qquad \text{initiation} \qquad (5.2)$$

$$CH_3\cdot + C_3H_8 \xrightarrow{k_2} CH_4 + C_3H_7\cdot \qquad \text{propagation} \qquad (5.9)$$

$$C_3H_7\cdot \xrightarrow{k_3} CH_3\cdot + C_2H_4 \qquad\qquad (5.8)$$

$$CH_3\cdot + CH_3\cdot \xrightarrow{k_4} C_2H_6 \qquad \text{chain ending} \qquad (5.12)$$

where $R_1 \equiv C_3H_7\cdot$, $R_2 \equiv CH_3\cdot$, $R_3 \equiv CH_3\cdot$, $R_4 \equiv C_3H_7\cdot$.

By application of the stationary state hypothesis to $CH_3\cdot$ and $C_3H_7\cdot$ we can write

$$d(CH_3\cdot)/dt$$
$$= k_1(C_3H_8) - k_2(CH_3\cdot)(C_3H_8) + k_3(C_3H_7\cdot) - 2k_4(CH_3\cdot)^2$$
$$= 0; \qquad (5.12)$$

$$d(C_3H_7\cdot)/dt = k_2(CH_3\cdot)(C_3H_8) - k_3(C_3H_7\cdot) = 0. \qquad (5.13)$$

Hence $(C_3H_7\cdot) = (k_2/k_3)(CH_3\cdot)(C_3H_8)$ and by substitution in (5.12),
$k_1(C_3H_8) - k_2(CH_3\cdot)(C_3H_8) + k_2(CH_3\cdot)(C_3H_8) - 2k_4(CH_3\cdot)^2 = 0.$

Solution for $(CH_3\cdot)$ gives

$$(CH_3\cdot) = \left\{ \frac{k_1}{2k_4}(C_3H_8) \right\}^{\frac{1}{2}}. \qquad (5.14)$$

The overall rate of decomposition has the form

$$\frac{-d(C_3H_8)}{dt} = k_1(C_3H_8) + k_2(CH_3\cdot)(C_3H_8). \qquad (5.15)$$

Now k_1 will be small relative to k_2 and hence, by substituting (5.14) in (5.15), we obtain the final rate expression

$$\frac{-d(C_3H_8)}{dt} = k_2 \left(\frac{k_1}{2k_4} \right)^{\frac{1}{2}} (C_3H_8)^{3/2}. \qquad (5.16)$$

In the case of ethane decomposition, it was stated that reaction type (5.5) was faster than type (5.4). Hence the chain ending step will involve recombination of $R_1\cdot$ radicals rather than $R_3\cdot$ as in propane decomposition, i.e. in the case of ethane decomposition, the recombination step will be

$$C_2H_5 \cdot + C_2H_5 \cdot \rightarrow C_4H_{10}. \qquad (5.17)$$

The classical Rice-Herzfeld mechanism for ethane decomposition is then written as follows:

$$C_2H_6 \xrightarrow{k_1} 2CH_3 \cdot \qquad (5.1)$$

$$CH_3 \cdot + C_2H_6 \xrightarrow{k_2} CH_4 + C_2H_5 \cdot \qquad (5.18)$$

$$C_2H_5 \cdot \xrightarrow{k_3} H \cdot + C_2H_4 \qquad (5.6)$$

$$H \cdot + C_2H_6 \xrightarrow{k_4} H_2 + C_2H_5 \cdot \qquad (5.7)$$

$$2C_2H_5 \cdot \xrightarrow{k_5} C_4H_{10}$$

where $R_1 \equiv C_2H_5\cdot$, $R_2 \equiv H\cdot$, $R_3 \equiv H\cdot$, $R_4 \equiv C_2H_5\cdot$.

Applying the stationary state hypothesis to $C_2H_5\cdot$, $H\cdot$ and $CH_3\cdot$

$$\frac{d(C_2H_5\cdot)}{dt} = k_2(CH_3\cdot)(C_2H_6) - k_3(C_2H_5\cdot) + k_4(H\cdot)(C_2H_6)$$
$$- 2k_5(C_2H_5\cdot)^2 = 0; \qquad (5.18)$$

$$\frac{d(H\cdot)}{dt} = k_3(C_2H_5\cdot) - k_4(H\cdot)(C_2H_6) = 0; \qquad (5.19)$$

$$\frac{d(CH_3\cdot)}{dt} = 2k_1(C_2H_6) - k_2(CH_3\cdot)(C_2H_6) = 0. \qquad (5.20)$$

From (5.20) and (5.19) we obtain

$$(CH_3\cdot) = \frac{2k_1}{k_2} \quad (5.21), \qquad (H\cdot) = \frac{k_3}{k_4}\frac{(C_2H_5\cdot)}{(C_2H_6)}. \qquad (5.22)$$

Substitution for $(CH_3\cdot)$ and $(H\cdot)$ in (5.18) results in

$$2k_1(C_2H_6) - k_3(C_2H_5\cdot) + k_3(C_2H_5\cdot) - 2k_5(C_2H_5\cdot)^2 = 0$$

which, by solution for $(C_2H_5\cdot)$, gives

$$(C_2H_5\cdot) = \left(\frac{k_1}{k_5}\right)^{\frac{1}{2}} (C_2H_6)^{\frac{1}{2}}. \qquad (5.23)$$

The overall rate equation is given by

$$\frac{-d(C_2H_6)}{dt} = k_1(C_2H_6)+k_2(CH_3\cdot)(C_2H_6)+k_4(H\cdot)(C_2H_6)$$

$$= k_1(C_2H_6)+2k_1(C_2H_6)+k_3(C_2H_5\cdot) \tag{5.24}$$

$$= k_1(C_2H_6)+2k_1(C_2H_6)+k_3\left(\frac{k_1}{k_5}\right)^{\frac{1}{2}}(C_2H_6)^{\frac{1}{2}} \tag{5.25}$$

by substitution for $(CH_3\cdot)$, $(H\cdot)$ and $(C_2H_5\cdot)$. Now k_1 will be much less than k_3 and hence the following rate equation results

$$\frac{-d(C_2H_6)}{dt} = k_3\left(\frac{k_1}{k_5}\right)^{\frac{1}{2}}(C_2H_6)^{\frac{1}{2}}. \tag{5.26}$$

The mechanisms which are given above are of course much simplified and one complication in the case of propane has already been referred to. Others will now be considered.

As stated previously, workers concerned with mechanistic elucidation have invariably carried out experiments to low degrees of conversion, perhaps to 1 or 2% at most, with the intention of eliminating secondary reactions involving products. However, it has been shown that reaction rates fall off at degrees of decomposition much below this, and this observation is interpreted in terms of self-inhibition in which the chain carriers are removed by reaction with olefinic products. Self-inhibition has been recognised for many years but mechanisms have rarely taken account of it. The phenomenon is obviously of great importance on the industrial scale where decomposition is carried to far higher levels than in the mechanistic studies.

In the case of ethane and propane pyrolysis the inhibition reactions are

$$H+C_2H_4 \rightarrow C_2H_5\cdot \quad \text{and} \quad H+C_3H_6 \rightarrow CH_3CH\cdot CH_3$$

In the ethane case the product distribution remains the same irrespective of the degree of self inhibition. In the case of propane again there is little substantial change in the product distribution except that as the propylene concentration increases, secondary propyl radicals will replace methyl as the principal reactants in the chain termination step resulting in different products.

Another complicating factor in hydrocarbon pyrolysis is the possible occurrence of surface reactions. This has been a widely investigated subject over the years, and there is still much disagreement on the importance of heterogeneous reactions. It used to be common practice to produce conditioned vessel surfaces by pre-pyrolysing hydrocarbons thereby depositing a carbonaceous layer, but it is now known that such layers in fact are extremely reactive. It is also generally accepted now that in the absence of carbonaceous layers and if oxygen, which has been shown to have an extreme effect on decomposition rates, is firmly excluded, the pyrolyses of propane, n-butane and other hydrocarbons where H atoms are not important as products, are essentially homogeneous at the usual pyrolysis temperatures and pressures. On the other hand, the thermal decompositions of ethane and isobutane have been shown to be markedly heterogeneous; here in fact carbonaceous layers are desirable for reproducible results. Increasing the surface/volume ratio results in a decrease in decomposition rate. To account for this observation it is suggested that H atoms, which are major products in ethane and isobutane pyrolyses, can diffuse to the wall quickly and there react, thus removing them from the chain propagating sequence. As was seen earlier the decomposition of propane can give rise to H atoms when the decomposition of 2-propyl radicals is involved. Surface effects have in fact been reported for propane at low temperatures and low pressures suggesting that under these conditions H atoms are formed and removed heterogeneously.

The result of these complications is reflected in a change in the order of reaction. Heterogeneous termination involving H atoms increases the overall order by up to $\frac{1}{2}$ unit, while the occurrence of self-inhibition has the effect of reducing the overall order.

The final complication to be discussed concerns the possible isomerisation of radicals. For example, in the case of propane decomposition, where 2-propyl may be in substantial amount, isomerisation to 1-propyl can take place.

$$CH_3CH \cdot CH_3 + C_3H_8 \rightarrow CH_3CH_2CH_2 \cdot + C_3H_8.$$

This reaction will be enhanced by high reactant pressures and thus the nature of the reaction products will change with pressure, methane being favoured relative to hydrogen at higher pressures.

At high olefin pressures other secondary reactions become important. For example, in ethane pyrolysis the following polymerisation/isomerisation/decomposition reaction sequence has been postulated:

$$C_2H_5{\cdot}+C_2H_4 \rightleftharpoons 1-C_4H_9{\cdot}$$

$$1-C_4H_9{\cdot}+C_2H_4 \rightleftharpoons 1-C_6H_{13}{\cdot}$$

$$1-C_6H_{13}{\cdot} \rightleftharpoons 2-C_6H_{13}{\cdot}$$

$$2-C_6H_{13}{\cdot} \rightarrow 1-C_3H_7{\cdot}+C_3H_6$$

$$1-C_3H_7{\cdot} \rightarrow CH_3{\cdot}+C_2H_4.$$

This type of reaction can lead to a wide variety of products.

(b) From the above discussion, which has been limited largely to consideration of experimental work carried out to low degrees of conversion, it is obviously extremely difficult to account for all the observations by a simple mechanism. Thus at the greater degrees of decomposition which are employed in actual commercial operation it is even more improbable that satisfactory mechanistic schemes can be devised because secondary reactions assume much greater importance.

As the conversion increases so the partial pressures of primary products increase. These can then either enter into inhibition reactions involving the addition of free-radicals, which have already been referred to, or themselves undergo cracking reactions with the eventual formation of acetylenes, diolefines and aromatics. It is observed that initially the primary product yields increase linearly with conversion and pass through the origin, but as conversion increases, deviation from linearity occurs, generally upwards for hydrogen and methane and down for propylene, butenes and other higher olefines as their cracking rates become significant. These curves can go through maxima. The ethylene curve goes either up or down depending on whether its rate of formation from the higher olefine pyrolysis is greater than or less than the rate of conversion of ethylene to secondary products. This behaviour will depend upon the type of hydrocarbon being cracked. In naphtha pyrolysis all reactions in the later stages will be secondary.

A plot of product yield versus conversion for propane cracking is shown in Fig. 5.1. Methane, hydrogen, ethylene and ethane yields all increase with conversion but that of propylene passes through a maximum. Cracking reactions take place in a tubular reactor with a space time of around a second or less. Fig. 5.2 is a product yield plot against a parameter related to the space time, namely the volume of reactor/molar feed rate of reactant. Since conversion will obviously increase with space time there is a distinct similarity in shape between Figs. 5.1 and 5.2.

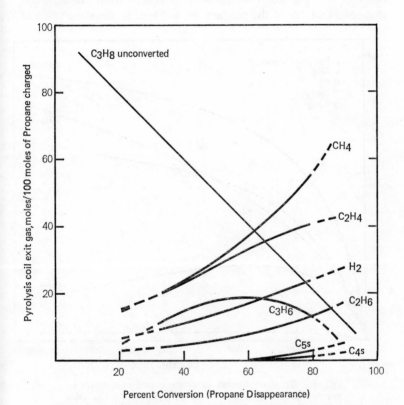

FIG. 5.1. Product distribution curves for propane pyrolysis (Reprinted with permission from *Chem. Eng. Prog.* **43**, 103 (1947))

It has been seen earlier in this section that it is extremely difficult to specify a precise order of reaction since this depends to a great extent on the nature of the termination reaction which does not always remain the same during reaction. Suppose to a first approximation that the order is first. Then the conversion will be related to residence time by equation (2.5):

$$kt = 2\cdot303 \quad \log \frac{1}{1-X_A}$$

where X_A is the fractional conversion of reactant and k is the rate constant at a particular temperature. It is seen from this equation that combinations of the product, kt, will lead to the same value of

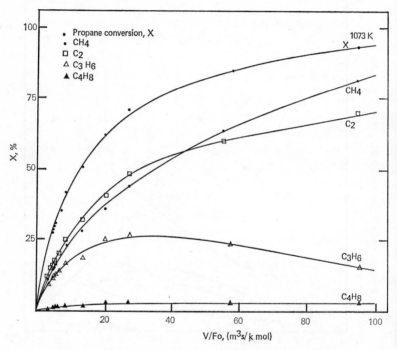

FIG. 5.2. Total conversion of propane and conversion to primary products as a function of V/F_0 at 1073°K (Reprinted from *Ind. Eng. Chem.* (*Proc. Des. & Dev.*), **7**, (1968), 435 Copyright 1968 by the American Chemical Society. Reprinted by permission of the copyright owner)

X_A, and thus high temperature of operation coupled with short reactor residence time will result in similar conversions to those obtained at low temperatures and longer residence times.

Although the conversion is seen to be a function of kt the product compositions will not be the same at equivalent values of kt since reactions other than the primary decomposition will depend on temperature to a different extent. For example, in naphtha cracking, the ethylene yield increased from 20–23% to 30–35% at the same conversion as the temperature was raised from 1020 to 1060°K and the residence time lowered correspondingly (Frank, 1969).

5.4 Conditions of Operation of Pyrolysis Reactors as Suggested by Analysis of Thermodynamic and Kinetic Features

It has been established in section 5.2 that cracking must be carried out at temperatures in excess of about 1000°K for satisfactory product yields. The product yield and distribution is determined by kinetic considerations, the important parameters being temperature, which determines the value of the rate constant, and reactor residence time; although as seen in section 5.3, the degree of decomposition is approximately proportional to the product of rate constant and residence time, final product distributions are determined by both these parameters individually. In particular ethylene formation is increased by operating at high temperatures and low residence times.

Experience has shown that it is desirable to carry out cracking reactions under conditions such that the primary reactions are allowed to attain a partial equilibrium but any approach to equilibrium with regard to side reactions is to be avoided. For example, in ethane cracking, it is desirable that the ethylene-ethane reaction attain a partial equilibrium, but not the reaction involving the removal of ethylene through further reaction with ethane. A useful control on the occurrence of secondary reactions is to ensure that the extent of partial equilibrium, or the degree of approach to equilibrium of the primary reaction in question, does not exceed an empirically determined value, the measure of which depends on the conditions of operation. The degree of approach is defined by K_{app}/K where K_{app} is the partial pressure ratio for the reaction under consideration, and K is the true equilibrium constant at the temperature in question. Thus in the case of the primary reaction

$$C_2H_6 \rightleftharpoons C_2H_4 + H_2,$$

K_{app} is given by
$$K_{app} = \frac{p_{C_2H_4} \, p_{H_2}}{p_{C_2H_6}}$$

where the partial pressures and temperature at which K is evaluated are measured at the same point in the pyrolysis tube. The degree of approach must not be higher than a certain empirically determined value for successful operation. This is especially so in the region of high conversion. Obviously since the equilibrium constant increases with rise in temperature the allowable ethylene and hydrogen partial pressures can increase, i.e. the allowable extent of conversion can increase with rise in temperature.

The approach to equilibrium is dependent upon the mass velocity, the intensity of heat application and the amount of steam dilution. Schutt (1961) has given a correlation for the equilibrium approach for propane decomposition as a function of conversion, heat flux and the steam dilution, viz. $A = K_{app}/K = \alpha - e^{\beta(X_A - \gamma)}$ where α, β and γ are constants and X_A is conversion (Table 5.2 and Fig. 5.3).

TABLE 5.2

Characteristic heat intensity in conversion zone W/m^2	Steam dilution mol/mol of propane feed	Value of coefficients		
		α	β	γ
31500	0·30−0·35	1·080	4·37	1·007
31500	0·55−0·60	1·085	4·00	1·043
63000	0·30−0·35	1·065	4·76	0·966
63000	0·55−0·60	1·070	4·41	0·991

The reason for the lower allowable degree of approach for higher heat fluxes is that under this condition the temperature gradient across the tube is large, the temperature next to the tube wall being high. The rates of the second order degradation reactions, which would be enhanced at high temperatures, must be reduced by suitable lowering of the partial pressures of these secondary reactants by restricting the conversion. Steam has a beneficial effect because its addition also serves to reduce the secondary reactant partial pressures. The effect of increasing mass flow rates not only increases the rate of heat transfer from the tube wall to the gas, but also reduces the thickness of the retarded layer of gas near the wall in which secondary reactions are particularly prone to occur, and which lead to coke formation. Of course high mass flow rates require high pressures to overcome pressure drop which in turn implies high partial pressures of hydrocarbons. However, these can be reduced by steam dilution.

A suggested mode of operation is to use heat fluxes as high as possible in the section of the pyrolysis tube just after the preheat

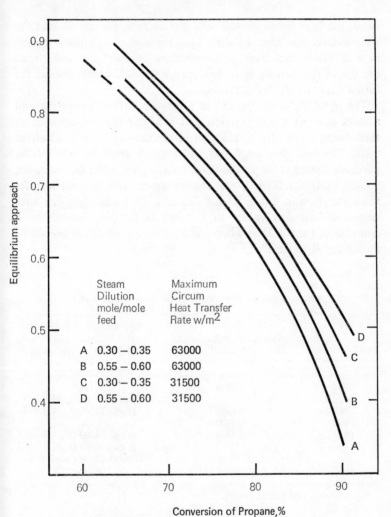

Steam
Dilution
mole/mole
feed

Maximum
Circum
Heat Transfer
Rate w/m^2

	Steam Dilution mole/mole feed	Maximum Circum Heat Transfer Rate w/m^2
A	0.30 – 0.35	63000
B	0.55 – 0.60	63000
C	0.30 – 0.35	31500
D	0.55 – 0.60	31500

Conversion of Propane,%

FIG. 5.3. Relation between conversion of propane and equilibrium approach

section. At this point, although the high heat flux requires a large temperature drop from the wall to the bulk of gas and although the 'wall gas temperature' will be high, this is not detrimental to successful operation because the initial reactants are still present in high enough concentrations for secondary reactions to be relatively unimportant. In later sections of the pyrolysis tube on the other hand, the heat fluxes should only just be high enough so that the temperature continues to rise. This method of operation results in a relatively high average temperature of operation and allows the use of reaction times of less than a second, which favours the formation of lower olefine products.

The pyrolysis is carried out in a furnace having convection and radiant sections, the convection part acting as the preheat section, there being essentially no reaction before the gas enters the radiant zone. The heat flux employed will depend upon the size of the pyrolysis tube, the bigger the diameter the greater the flux since the volume/surface ratio is higher. Some typical average heat transfer rates are given in Table 5.3 for different sized tubes together with suggested heat flux values to be used in the zone of advanced conversion, i.e. the zone where the equilibrium approach should be most carefully controlled.

TABLE 5.3

Effect of reactor tube diameter on allowed heat flux *

Pyrolysis tube O.D., mm.	Average heat flux in coil section ($> 820°K$), W/m² outside surface area	Approximate value of heat flux in critical reaction zone, W/m² outside surface area
50·8	28400	—
63·5	—	14800–20500 (C_2H_6)
76·2	31500	15800–23700 (C_2H_6)
88·9	33100	—
101·6	37900	12600–20500 (C_3H_8)
		22100–31500 (C_2H_6)
		9500–12600 (Naphtha)
114·3	39400	—

* *From* Snow and Schutt (1957).

For the same feed rate, increasing the heat flux and decreasing the gas residence time at the same conversion increases the yield of hydrogen, ethylene and propylene and outlet gas temperature and decreases the yield of ethane and secondary products and coke build-up. For the same conditions, but increasing the feed conversion by raising the total heat input, the yields of all major components (except propylene) increase but outlet temperatures and coke build up also increase.

In summary it has been deduced that the following conditions are desirable for the operation of cracking reactors for ethylene production:

(1) Ethylene yields are improved by operating at as high temperatures and as low residence times as possible ($\sim 1100°$K, and 0·6–1·3 s).

(2) High temperature operation implies high heat fluxes which in the later stages of reaction tends to result in secondary product and coke formation. This tendency is reduced by introduction of steam diluent into the feed; the steam lowers the partial pressures of hydrocarbons, reduces the residence time and keeps the furnace tubes clean. High flow rates also reduce secondary product formation.

(3) The equilibrium approach value for the cracking operation being considered should not be allowed to exceed that calculated for the conditions of operation being employed.

5.5 Kinetic Models for Hydrocarbon Pyrolysis Reactions

Ideally, an accurate kinetic model should predict the correct overall order and activation energy for the decomposition process and should also give satisfactory values for product yields at all stages of the reaction. In the section dealing with the kinetics of hydrocarbon pyrolysis reactions it was seen that, even at low degrees of conversion when secondary reactions were comparatively unimportant, a satisfactory mechanism which took account of all the probable reactions and which correctly predicted the order for the overall reaction was still a goal not yet achieved. Nevertheless rate equations derived according to this mechanistic approach are obviously the most appealing since they are based on a proper appreciation of the actual reaction steps involved.

Snow *et al.* (1959) employed this approach to predict expected product yields for ethane decomposition. An improved mechanism of the Rice-Herzfeld type was adopted, the reverse of some of the reactions postulated in the kinetic section and additional radical-radical reactions being included. By applying the steady state assumption, expressions were written down giving the concentrations of the active species, $CH_3\cdot$, $C_2H_5\cdot$ and $H\cdot$. These expressions were then substituted into equations for the formation rates of products. By multiplying these rates by a small interval of time the change in concentration of each component was obtained. New arithmetic averages of (C_2H_6), (CH_4), (C_2H_4) and (H_2) were computed and the free radical concentrations and reaction rates recalculated. The procedure was repeated until no further change in (C_2H_6) occurred. New intervals of time were then chosen and the procedure repeated until the correct conversion was obtained. For a flow system the procedure is similar, this time the variable is distance along reactor rather than time and correction is made for volumetric expansion of gases. With some adjustment of the values of the published rate constants for some of the reaction steps, a satisfactory fit was obtained to the available experimental data. This model still did not take account fully of the secondary reactions involved but the introduction of some of the back reaction steps was considered to be an improvement over the simple Rice-Herzfeld derivation. The approach just described of course is only possible through the use of a computer in the iterative calculations involved.

The more usual approach to the development of a kinetic model has been to postulate a number of empirical rate equations based on an analysis of product distributions over a wide range of reactor temperatures, residence times and degrees of conversion. Reactions are presented so that all observed products are accounted for. For example in ethane decomposition the following empirical equations have been postulated to account for the observed distributions (Snow and Schutt, 1957):

(a) $C_2H_6 \rightleftharpoons C_2H_4 + H_2$

$$r_1 = \frac{k_1 P}{n_T} \left(n_{C_2H_6} - \frac{n_{C_2H_4} \, n_{H_2} P}{K_1 n_T} \right)$$

(b) $C_2H_4 + 2H_2 \rightleftharpoons 2CH_4$

$$r_2 = \frac{k_2 P}{n_T} \left(\frac{P}{n_T} n_{C_2H_4} (n_{C_2H_6} \, n_{H_2})^{\frac{1}{2}} - \frac{n_{C_2H_4}}{K_2} \right)$$

(c) $C_2H_4 \rightarrow 0.25\,C_2H_6 + 0.125\,C_4H_8 + 0.125\,C_4H_{10} + 0.125\,H_2$

$$r_3 = 0.012\,r_1 P$$

(d) $C_2H_4 \rightarrow 0.333\,C_6H_6 + H_2$

$$r_4 = \frac{k_4 P^2}{n_T^2} \, n_{C_2H_4}^2$$

(e) $C_2H_4 \rightarrow C_2H_2 + H_2$

$$r_5 = \frac{k_5 P}{n_T^2} \, n_{C_2H_4}^2$$

(f) $C_2H_4 \rightarrow 2C + 2H_2$

$$r_6 = \frac{k_6 P^2}{n_T^2} \, n_{C_2H_4}^2$$

(g) $C_2H_4 + C_2H_6 \rightarrow 0 \cdot 952 \, C_3H_6 + 0 \cdot 381 \, C_3H_8 + 0 \cdot 62 \, H_2$

$$r_7 = \frac{k_7 P}{n_T} \left(n_{C_2H_6} - \frac{n_{C_2H_4} \, n_{H_2} P}{k_1 n_T} \right)$$

where n_i is number of moles of component i, n_T is total moles/mole of feed, P is total pressure. The rates of each of these reactions were calculated for small intervals of conversion, Δn_i, from the expression $r_i = \Delta n_i / V \Delta t$ and values of k for each reaction determined. V is the volume per mole of feed at the temperature and pressure of reaction and Δt is the time increment. The values of the rate constant should be the same at each stage of conversion for the reaction in question and a certain amount of trial and error fitting of the rate constant values was necessary before the calculated and experimental product yields agreed.

It should be emphasised that this form of model, being empirical, will only hold accurately over the range of reaction conditions on which the data are based.

5.6 The Use of Kinetic Models in the Design of Pyrolysis Reactors

Design of a pyrolysis reactor requires that for any set of reactor dimensions and applied heat flux the conversion, product distribution, temperature profile and pressure gradient in the pyrolysis tube shall be accurately predicted for a given feed stock. The usual design procedures involve the step-wise trial and error calculation of concentration changes of each product across each small increment of tube volume using empirical rate equations of the type presented in section 5.5. The temperature change across the increment should match that expected on the basis of degree of cracking in the increment and on the sensible heat change for the appropriate product distribution. Thus a trial and error method is necessary for accurate representation.

A simplified approach to the design of a tubular cracking reactor will now be considered. Instead of calculating the concentration profile for each product, the aim of the calculation will be to calculate the number of pyrolysis tubes of a specific size, i.e. the reactor volume necessary to achieve a given degree of conversion and which will reproduce the measured experimental product distribution at each point in the reactor. The temperature and pressure profiles along the reactor will also be deduced. For this calculation it is therefore necessary to have information on the product yields at various stages of conversion. For the case of propane cracking, which will be discussed, product distribution curves were shown in Fig. 5.1.

Suppose that the reaction may be considered first order. Then the reaction rate is given by

$$(-r_A) = kp_A = k \frac{n_A}{n_T} P \qquad (5.27)$$

where p_A is the partial pressure of reactant A, n_A the number of moles of A, n_T the total number of moles and P the total pressure at any point in the reactor.

Now $n_A = n_{A_0}(1 - X_A)$ and, if the rate is based on one mole of A initially present, the rate expression is given by

$$(-r_A) = \frac{kP(1 - X_A)}{n_T}. \qquad (5.28)$$

Writing $n_T = 1 + n_I + \delta X_A$ where n_I is number of moles of inert and δ is the expansion factor which accounts for the increase in the number of moles during reaction we obtain

$$(-r_A) = \frac{kP(1 - X_A)}{1 + n_I + \delta X_A}. \qquad (5.29)$$

By equation (2.34) for a tubular plug flow reactor

$$\frac{dV}{F_{A_0}} = \frac{dX_A}{(-r_A)}.$$

Substitution of equation (5.29) then leads to

$$\frac{dV}{dX_A} = \frac{F_{A_0}(1 + n_I + \delta X_A)}{kP(1 - X_A)}. \qquad (5.30)$$

This equation can be formally integrated to give

$$\frac{V}{F_{A_0}} = \int_{X_{A_1}}^{X_{A_2}} \frac{(1+n_1+\delta X_A)}{kP(1-X_A)} \, dX_A. \tag{5.31}$$

For a small increment of volume which, for instance, could be the volume of one reactor tube, k, P and δ may be considered constant, and equation (5.31) can be written as

$$\frac{k P \Delta V}{F_{A_0}} = \int_{X_{A_1}}^{X_{A_2}} (1+n_1) \frac{dX_A}{1-X_A} + \delta \int_{X_{A_1}}^{X_{A_2}} \frac{X_A}{1-X_A} \, dX_A.$$

On integration the resultant expression is

$$\frac{k P \Delta V}{F_{A_0}} = (1+n_1+\delta) \ln\left(\frac{1-X_{A_1}}{1-X_{A_2}}\right) - \delta(X_{A_2} - X_{A_1}). \tag{5.32}$$

For ease in computation the logarithmic term may be removed. Thus $\ln[(1-X_{A_1})/(1-X_{A_2})]$ may be replaced by $\ln[1 - \{(\Delta X_A)/(1-X_{A_2})\}]$ and, if ΔX_A (i.e. $X_{A_2} - X_{A_1}$) is small, series expansion of $\ln[1 - \{(\Delta X_A)/(1-X_{A_2})\}]$ gives $\Delta X_A/(1-X_{A_2})$.

Hence equation (5.32) can be written as

$$\frac{k P \Delta V}{F_{A_0}} = (1+n_1+\delta)\frac{\Delta X_A}{1-X_{A_2}} - \delta\Delta X_A$$

$$= \frac{\Delta X_A}{1-X_{A_2}} \, (1+n_1+\delta X_{A_2}). \tag{5.33}$$

Further, in regions where X_A is very small, (< 0.01), equation (5.33) simplifies to

$$\frac{k P \Delta V}{F_{A_0}} = (1+n_1) \, \Delta X_A. \tag{5.34}$$

Equations (5.33) and (5.34) were used by Fair and Rase (1954) in a trial and error method to calculate the number of tubes required to obtain a given degree of conversion. The procedure involves an initial assumption of the average temperature and pressure in the first tube and return bend (considered as the incremental volume, ΔV) in the radiant zone of the reactor. The rate constant at this temperature is then determined and X_{A_2} is calculated using either equation (5.34) if the conversion is < 0.01 or equation (5.33) if

> 0.01. ΔV, the expansion factor, δ, and the feed rate of reactant, F_{A_0}, must be known. For this first tube X_{A_1} is zero.

The assumed average temperature is then checked by carrying out a heat balance over the tube concerned and comparing the calculated temperature increase in the increment with that assumed, bearing in mind that the heat supplied to the tube is used for two purposes (1) cracking (2) sensible heating. If Δq is the heat supplied per tube, the heat of cracking, Δq_c will be given by

$$\Delta q_c = (\Delta X_A)(\Delta H)(F_{A_0}) \tag{5.35}$$

where ΔH is the heat of reaction at the temperature and conversion concerned. The rise in temperature in the increment is then determined from

$$\Delta T = \frac{\Delta q - \Delta q_c}{C_p \, F_0} \tag{5.36}$$

where C_p is the heat capacity of the reaction mixture at the appropriate conditions and F_0 is the total initial flow rate.

The rise in temperature along the tube, ΔT, should correspond to the assumed average tube temperature; if not, a new temperature must be assumed and the calculation repeated.

Likewise the assumed pressure in the tube must agree with the decrease in pressure along the tube calculated from

$$-\Delta P = \frac{(\text{p.d. factor})(\Delta L)}{P_{av}} \tag{5.37}$$

where P_{av} is the assumed average pressure in the tube, ΔL is the equivalent length of tube exposed to radiation and the pressure drop factor is calculated at the relevant temperature and conversion. Calculations are then performed for each increment until the required conversion is obtained.

Although this technique gives accurate results, the trial and error procedure makes computation lengthy. A method which removes the need for trial and error techniques when applied to the reaction zone, has been evolved by Perkins and Rase (1956) and this will be illustrated by an example in this section.

Equation (5.30) can be rearranged to give

$$\frac{dX_A}{dV} = \frac{kP(1 - X_A)}{F_{A_0}(1 + n_i + \delta X_A)}. \tag{5.38}$$

The typical shape of a plot of dX_A/dV vs. tube number is shown in Fig 5.4. For the straight line portion of the curve, $(dX_A/dV)_{av}$ in a tube is equal to (dX_A/dV) at the mid-way point of the tube. In addition (dX_A/dV) at the mid-way point will be equal to (dX_A/dV)

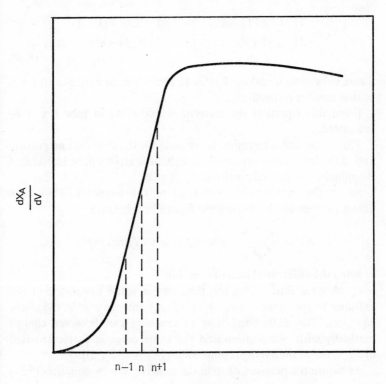

$n-1$ n $n+1$

Tube Number
(N.B. no. refers to tube entrance)

Fig. 5.4. Typical behaviour of conversion change (Reprinted with permission from *Chem. Eng. Prog.* **52**, 105-M (1956))

at the entrance of the tube plus half the change in (dX_A/dV) for the previous tube,

i.e. $(dX_A/dV)_{av} = (dX_A/dV)_n + \frac{1}{2}[(dX_A/dV)_n - (dX_A/dV)_{n-1}]$

in tube n $= \frac{3}{2}[dX_A/dV]_n - \frac{1}{2}[dX_A/dV]_{n-1}.$ (5.39)

Now since $(dX_A/dV)_{av} = (\Delta X_A/\Delta V)$ for the straight line portion

and $\left(\dfrac{dX_A}{dV}\right)_n = \left[\dfrac{kP(1-X_A)}{F_{A_0}(1+n_I+\delta X_A)}\right]_n,$ $\qquad$ (5.38)

then

$$\Delta X_A = \frac{3}{2}\left[\frac{kP\Delta V(1-X_A)}{F_{A_0}(1+n_I+\delta X_A)}\right]_n \quad -\frac{1}{2}\left[\frac{kP\Delta V(1-X_A)}{F_{A_0}(1+n_I+\delta X_A)}\right]_{n-1}$$
$$\tag{5.40}$$

where subscripts n and $n-1$ refer to conditions at entrance to tubes of that number respectively.

From this equation the conversion occurring in tube n can be calculated.

This numerical integration is accurate for the straight line portion and, if the increments are small enough, little error will be introduced if applied at the curved portions.

As in the more rigorous procedure the temperature increase across the tube can be determined from a heat balance

$$\Delta T = \frac{\Delta q - \Delta q_c}{C_p F_0} \quad \text{where } \Delta q_c = (\Delta X_A)(F_{A_0})(H),$$

H being the differential heat of cracking.

C_p changes little along the tube and a value calculated at the entrance temperature to tube $n-1$, (T_{n-1}), may be used in the above equation. The differential heat of cracking can however change markedly with composition and the value taken is that calculated at T_{n-1} and at the average composition $[X_{A_n} - (\Delta X_A/2)]$.

As before the pressure drop in the tube is given by equation (5.37)

$$-\Delta P = \frac{(\text{p.d. factor})(\Delta L)}{P_{av}}$$

where

$$-\Delta P = P_{n-1} - P_n,$$

and P_{av} is given by

$$\frac{P_n + P_{n-1}}{2}.$$

P_n is then deduced to be

$$P_n = [(P_{n-1})^2 - 2(\text{p.d. factor})(\Delta L)]^{\frac{1}{2}}. \tag{5.41}$$

This is evaluated at T_n and $[X_{A_n} - (\Delta X_A/2)]$.

This technique of Perkins and Rase can be applied without trial and error to the reactor section from the point at which the change in conversion with respect to reactor volume is $0.001/\Delta V$ if the temperature and pressure at this point are known (the original paper also gives a method of calculation for the number of tubes in the preheat section).

Example 5.1. Propane is fed at a rate of 3170 kg/h into a reactor consisting of a series of horizontal tubes connected by return bends. The tubes are made of stainless steel, 0.102 m in diameter, 7.32 m long, located on 0.305 m centres giving a tube volume of 0.06344m^3, heated outside surface area of 2.787m^2 and equivalent pressure drop length of 13.42 m. All tubes and return bends are exposed to a heat flux of 31.54 kW/m^2 outside area. Suppose that the temperature at the end of the preheat section is 871°K and the pressure 3.926 bar. Calculate the number of tubes, exit temperature and pressure for a conversion of 50%. (1 bar $\equiv 10^5$ N/m^2).

Solution. The solution of this problem is carried out by the numerical integration technique outlined earlier in this section. The following charts are required in the solution (1) Rate constant versus temperature (2) Heat capacity versus temperature (3) Differential heat of cracking versus % conversion (4) Pressure drop factor versus temperature and conversion. These are shown as Figs 5.5–5.8 respectively.

It is also useful in the calculation to have a plot of the function $[(1 - X_A)/(1 + \delta X_A)]$ vs. X_A. This is shown in Fig. 5.9.

The method of calculation will be illustrated for the first three tubes of the series.

Tube 1. The first tube will be that in which a conversion of 0.001 occurs. The temperature at the entrance to this tube is that at the end of the so-called preheat section, $T_1 = T_p$, which is 871°K, and the pressure, P_1, is 3.926 bar.

Tube 2. (a) Applying equation (5.36) the temperature at the entrance to the second tube is given by

$$T_2 = T_1 + \left(\frac{\Delta q - \Delta q_c}{C_p F_0}\right),$$

where $\Delta q_c = (\Delta X_A)(F_{A_0})(H)$
and Δq is heat incident on tube.

Since pure propane is being fed

$$F_0 = F_{A_0} = 3170 \text{ kg/h}$$

$$= \frac{3170}{3600 \times 44 \cdot 1} = 0 \cdot 0200 \text{ kmol/s}.$$

ΔX_A in the first tube is $0 \cdot 001$ and, from Fig. 5.7, H, calculated at T_1 and $[X_{A_2} - (\Delta X_A/2)] = 0 \cdot 0005$, is 86004 kJ/kmol.

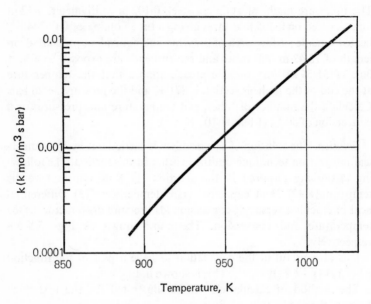

Fig. 5.5. Dependence of reaction velocity constant on temperature for propane decomposition (Reprinted with permission from *Chem. Eng. Prog.* **52**, 105-M (1956))

C_p is calculated at T_1 and has the value $164 \cdot 0$ kJ/°K (kmol propane fed). Heat incident on tube has the value

$$\Delta q = 31 \cdot 54 \times 2 \cdot 787 \text{ kJ/s}.$$

Hence substituting in (5.36)

$$T_2 = 871 + \frac{(31 \cdot 54 \times 2 \cdot 787) - (0 \cdot 001)(0 \cdot 0200)(86004)}{164 \cdot 0 \times 0 \cdot 0200}$$

$$= 897 \cdot 3°\text{K}.$$

(b) From Fig. 5.5, at this temperature, $k = 2 \cdot 16 \times 10^{-4}$ kmol/s m^3 bar.

(c) P_2 is calculated from equation (5.41)
$$P_2 = [(P_1)^2 - 2(\text{p.d. factor})(\Delta L)]^{\frac{1}{2}}.$$

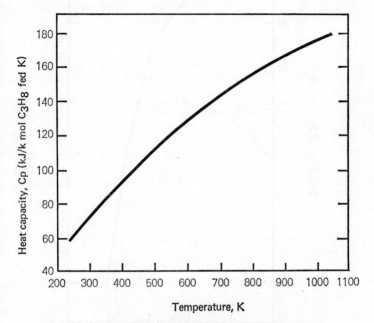

FIG. 5.6. Temperature dependence of heat capacity (based on propane fed) (Reprinted with permission—see Fig. 5.5)

The pressure drop factor is calculated using Fig. 5.8, calculation being made at temperature T_2 and $[X_{A_2} - (\Delta X_A/2)]$.
From Fig. 5.8 the factor is $0 \cdot 0180$.

Hence $P_2 = [(3 \cdot 926)^2 - 2(0 \cdot 0180)(13 \cdot 42)]^{\frac{1}{2}}$

$\qquad = 3 \cdot 864$ bar.

(d) From Fig. 5.9, $[(1 - X_A)/(1 + \delta X_A)]$ at $X_A = 0 \cdot 001$ is unity.

(e) The conversion occurring in tube 2 is then given by equation (5.40)
$$\Delta X_A = \frac{3}{2} \left[\frac{kP \, \Delta V (1 - X_A)}{F_{A_0}(1 + \delta X_A)} \right]_2 - \frac{1}{2} \left[\frac{kP \, \Delta V (1 - X_A)}{F_{A_0}(1 + \delta X_A)} \right]_1$$

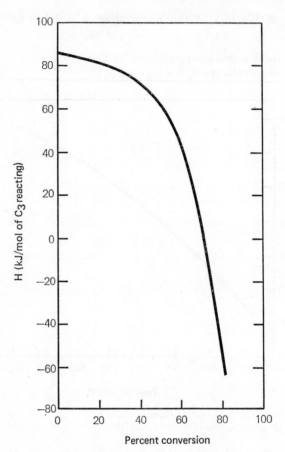

FIG. 5.7. Differential heat of cracking of propane (800–1100°K) (Reprinted with permission—see Fig. 5.5)

$$= \frac{3}{2}\left[2 \cdot 16 \times 10^{-4} \times 3 \cdot 864 \times \frac{0 \cdot 06344}{0 \cdot 020} \times 1\right] \quad -\frac{1}{2}[0 \cdot 001]$$

$$= \tfrac{3}{2}(0 \cdot 00264) - \tfrac{1}{2}(0 \cdot 001) = 0 \cdot 00346.$$

Hence total conversion at exit of tube 2 is $0 \cdot 00346 + 0 \cdot 001 = 0 \cdot 00446$.

Tube 3. (a) $T_2 = 897 \cdot 3°$K.

$$\left[X_{A_2} - \frac{\Delta X_A}{2}\right] = 0 \cdot 00446 - \frac{0 \cdot 00346}{2} = 0 \cdot 00273.$$

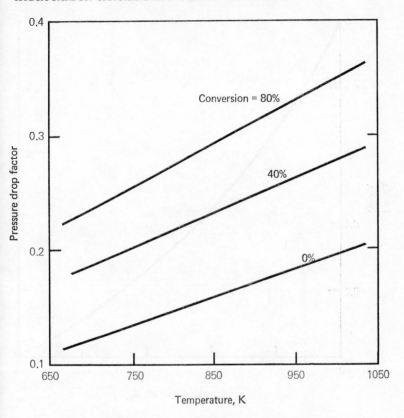

FIG. 5.8. Pressure drop factor for propane cracking
(Reprinted with permission—see Fig. 5.5)

C_p is evaluated at T_2 and H at T_2 and $[X_{A_2} - (\Delta X_A/2)]$.

T_3 is then evaluated from equation (5.36):

$$T_3 = T_2 + \left(\frac{\Delta q - \Delta q_c}{C_p F_0} \right)$$

$$= 897 \cdot 3 + \frac{(31 \cdot 54)(2 \cdot 787) - (0 \cdot 00346)(0 \cdot 02)(86004)}{0 \cdot 02 \times 166 \cdot 52}$$

$$= 922 \cdot 2°K.$$

F

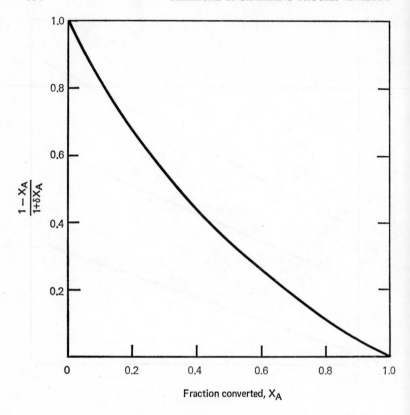

FIG. 5.9. Conversion and expansion factor for propane
(Reprinted with permission—see Fig. 5.5)

(b) k at $922 \cdot 2°K = 5 \cdot 07 \times 10^{-4}$ kmol/s m³ bar.

(c) $P_3 = [(P_2)^2 - 2 \text{ (p.d. factor)}(\Delta L)]^{\frac{1}{2}}$

$\quad\quad = [(3 \cdot 864)^2 - 2(0 \cdot 0184)(13 \cdot 42)]^{\frac{1}{2}}$

$\quad\quad = 3 \cdot 802$ bar. The p.d. factor is calculated at T_3 and

$\quad\quad [X_{A_3} - (\Delta X_A/2)].$

(d) From Fig. 5.9 $(1 - X_A)/(1 + \delta X_A)$ is unity.

(e) Conversion occurring in tube 3 is given by equation (5.40)

$$\Delta X_A = \frac{3}{2}\left[\frac{kP\,\Delta V(1-X_A)}{F_{A_0}(1+\delta X_A)}\right]_3 - \frac{1}{2}\left[\frac{kP\,\Delta V(1-X_A)}{F_{A_0}(1+\delta X_A)}\right]_2$$

$$= \frac{3}{2}\left[5{\cdot}07\times10^{-4}\times3{\cdot}80\times\frac{0{\cdot}0634}{0{\cdot}02}\times1\right] - \frac{1}{2}[0{\cdot}00264]$$

$$= \tfrac{3}{2}(0{\cdot}00611) - \tfrac{1}{2}(0{\cdot}00264) = 0{\cdot}00784.$$

Hence total conversion at exit of tube 3 is $0{\cdot}00784+0{\cdot}00446 = 0{\cdot}01230$.

Calculations are repeated in an analogous fashion to that for tube 3 until the desired conversion is reached. The full results are shown in Table 5.4.

It is seen that 16 tubes are required to achieve a conversion of 50%. There are limitations on the accuracy of these calculations because of the errors inherent in reading the graphs. The accuracy would obviously be improved by improved data.

TABLE 5.4

Tube No.	Temperature at entrance to tube, °K	Pressure at entrance to tube, bar	ΔX_A	% Conversion at exit of tube
1	871	3·926	0·001	0·1
2	897	3·864	0·00346	0·446
3	922	3·802	0·00784	1·230
4	944	3·733	0·0155	2·78
5	962	3·664	0·0248	5·26
6	975	3·588	0·0327	8·53
7	985	3·512	0·0385	12·38
8	991	3·429	0·0404	16·42
9	996	3·333	0·0409	20·51
10	1002	3·243	0·0433	24·84
11	1007	3·139	0·0425	29·09
12	1011	3·022	0·0442	33·51
13	1016	2·905	0·0418	37·69
14	1023	2·774	0·0468	42·30
15	1028·5	2·594	0·0444	46·74
16	1036	2·470	0·0450	51·24

REFERENCES

FAIR, J. R. and RASE, H. F. 1954. *Chem. Eng. Prog.* **50**, 415.

FRANK, S. M. 1969. In *Ethylene and its industrial derivatives*, ed. Miller, S. A. Benn, London, Chapter 3, p. 131.

LEATHARD, D. A. and PURNELL, J. H. 1970. *Ann. Rev. Phys. Chem.* **21**, 197.

PERKINS, R. K. and RASE, H. F. 1956. *Chem. Eng. Prog.* **52**, 105–M.

RICE, F. O. and HERZFELD, K. F. 1934. *J. Amer. Chem. Soc.* **56**, 284.

SCHUTT, H. C. 1961. *Z. Electrochem.* **64**, 245.

SNOW, R. H., PECK, R. E., and VON FREDDERSDORF, C. G. 1959. *A.I.Ch.E.J.* **5**, 304.

SNOW, R. H. and SCHUTT, H. C. 1957. *Chem. Eng. Prog.* **53**, 133–M.

6

MULTI-STEP REACTION PROCESSES

6.1 Introduction

An ideal chemical process would be one where product formation occurs at a high rate and where by-product formation is negligible. Selectivity of this degree is generally obtained by correct choice of catalyst for the process; the synthesis of ammonia over an iron catalyst is an example of a reaction which is unaccompanied by side-reactions. The cracking of saturated hydrocarbons is an example of a reaction which is extremely complex and where the main desirable product, ethylene, cannot be obtained other than in admixture with other products. Nevertheless, despite this complexity, it has been shown in Chapter 5 that a simple first order rate equation may be used to obtain a preliminary design for a pyrolysis reactor provided that experimental data on product distributions are available.

Between these two extremes of processes involving no by-product formation and those where the extent of side reactions is very marked, and hence where control of reaction is difficult, there are certain cases where competing reactions are amenable to mathematical description and where it is readily possible to predict conditions for optimum production rate of the desired product. There are three general classes of complex reactions:

(a) consecutive reactions of the type

$$A \rightarrow B \rightarrow C \rightarrow$$

where B is the required product.

(b) parallel reactions of the type

$$A \rightarrow B$$
$$A \rightarrow C$$

where B is the required product and C is a by-product.

(c) parallel—consecutive reactions of the type

$$A+B \rightarrow C$$

$$C+B \rightarrow D$$

$$D+B \rightarrow E \qquad \text{etc.}$$

where the reaction is parallel with respect to reactant B and consecutive with respect to A, C, D and E.

In this chapter examples of the type (a) and (c) will be considered, acetylene formation from hydrocarbons being considered as an example of type (a) and chlorination of hydrocarbons as an example of type (c).

6.2 Acetylene Production from Hydrocarbons

Because the acetylene molecule contains a triple bond, addition reactions take place readily. In principle, therefore, acetylene could be a most important raw material in organic chemical manufacture. This is indeed the case, and acetylene has found use as a starting material in the synthesis of large tonnage chemicals including vinyl chloride, acrylonitrile and vinyl acetate. The utility as a raw material depends largely on economic factors and synthesis from acetylene has not grown to the extent that it might because the products mentioned can equally well be produced from the less costly olefines, ethylene or propylene.

In the past most of the acetylene has been produced through the carbide route which in fact is probably still the most important source. More recent developments have been based on a route through hydrocarbons and it is processes based on this route which will be considered here.

6.2.1 THERMODYNAMICS OF ACETYLENE PRODUCTION FROM HYDROCARBONS

Acetylene can be produced by pyrolysis of methane (natural gas), refinery off gases (ethane, propane) or naphtha. Examination of the standard free energy changes occurring in the reactions will indicate the temperature conditions which must be used in the process. A plot of standard free energy of formation for low carbon number hydrocarbons against temperature is shown in Fig. 6.1. Acetylene

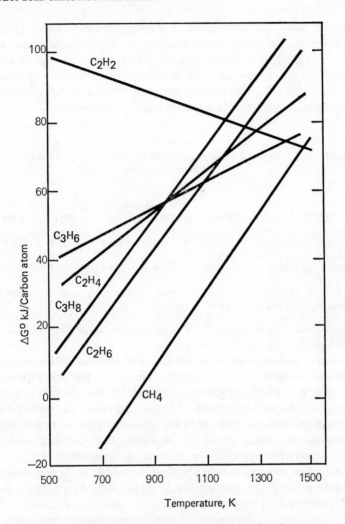

FIG. 6.1. Standard free energy of formation of hydrocarbons as a function of temperature

is seen to be unstable with respect to other hydrocarbons at low temperatures, but the stability increases with temperature until at temperatures above 1500°K it becomes the most stable of hydrocarbons. Free energy changes for the conversion of methane and ethane to acetylene are shown in Table 6.1. If reaction is considered feasible when $\Delta G° =$ zero then a threshold temperature of $1225°K$ is necessary for ethane decomposition, but for methane a temperature $> 1485°K$ is required. Also listed in Table 6.1 are the standard free energy change for the acetylene decomposition reaction, $C_2H_2 \rightleftharpoons 2C + H_2$, and the standard heat changes in these reactions. The formation reaction is seen to be highly endothermic.

TABLE 6.1

Standard free energy and enthalpy changes for acetylene synthesis reactions from methane and ethane

	Temperature, °K				
$(\Delta G°)_r$, kJ/mol	1000	1200	1300	1400	1500
$2CH_4 = C_2H_2 + 3H_2$	133·2	76·9	49·6	22·4	−4·9
$C_2H_6 = C_2H_2 + 2H_2$	60·6	6·9	−19·9	−46·7	−73·6
$C_2H_2 = 2C + H_2$	−169·9	−159·4	−159·4	−149·1	−144·0
$(\Delta H°)_r$, kJ/mol					
$2CH_4 = C_2H_2 + 3H_2$	402·3	404·1	404·6	404·6	404·5
$C_2H_6 = C_2H_2 + 2H_2$	328·8	329·0	328·7	328·2	327·5
$C_2H_2 = 2C + H_2$	−223·0	−221·8	−221·1	−220·5	−219·9

At the temperature at which the acetylene formation reaction becomes feasible, the free energy change for the decomposition reaction is markedly negative, indicating that this reaction is highly favoured thermodynamically. Thus, if the reaction system were allowed to reach equilibrium at the calculated process temperature, acetylene formation would be negligible and decomposition to carbon and hydrogen would be substantially complete.

A further conclusion from the thermodynamic analysis is that the acetylene formation reaction would be favoured by operating at a low pressure or in the presence of inert material, since reaction occurs with an increase in the number of product over reactant molecules.

6.2.2 KINETICS OF ACETYLENE FORMATION.

Since acetylene yields would obviously be extremely small if equilibrium conditions were attained, it is necessary, if possible, to choose a process time such that the formation reaction proceeds to a satisfactory extent, but the decomposition reaction does not become unduly important. A simplified reaction scheme of the following type could be written to describe acetylene formation:

Saturated hydrocarbon → acetylene → carbon + hydrogen.

Little information exists on the kinetics of decomposition of hydro-carbons other than methane, and even here the exact kinetics of reaction are not known. There is evidence to suggest that ethylene is present as an intermediate in the partial oxidation process (described later) for acetylene production. In fact it is probable that ethane is a precursor of the ethylene, but the rate of formation of ethylene from ethane is so fast that ethane would be present in very small amounts in the products at any time. Therefore a satisfactory reaction scheme could involve the following steps:

$$CH_4 \xrightarrow{k_1} C_2H_4 \xrightarrow{k_2} C_2H_2 \xrightarrow{k_3} \text{products} \tag{6.1}$$

where k_1, k_2 and k_3 are first order rate constants.

Now a first order reaction scheme of this type involving two re-action steps has been considered in Chapter 2. By similar procedures to those developed in the earlier chapter, rate equations can be written down for a three step process and the equations integrated under isothermal conditions to produce expressions for each of the inter-mediates as a function of time. For reaction (6.1) the following rate equations can be formulated:

$$d(CH_4)/dt = -k_1(CH_4) \tag{6.2}$$

$$d(C_2H_4)/dt = k_1(CH_4) - k_2(C_2H_4) \tag{6.3}$$

$$d(C_2H_2)/dt = k_2(C_2H_4) - k_3(C_2H_2). \tag{6.4}$$

Integration of (6.2) and (6.3) leads to (see section 2.2.4)

$$(CH_4) = (CH_4)_0 \, e^{-k_1 t}, \tag{6.5}$$

and

$$(C_2H_4) = (CH_4)_0 \, \frac{k_1}{k_2 - k_1} \{e^{-k_1 t} - e^{-k_2 t}\} \tag{6.6}$$

where $(CH_4)_0$ is initial concentration of methane. Substitution in equation (6.4) for ethylene concentration results in

$$\frac{d(C_2H_2)}{dt} = k_2(CH_4)_0 \, \frac{k_1}{k_2 - k_1} \{e^{-k_1 t} - e^{-k_2 t}\} - k_3(C_2H_2). \tag{6.7}$$

This linear differential equation can be solved using the factor

$e^{k_3 t}$ and the constant evaluated from the condition that at $t = 0$, $(C_2H_2)_0 = 0$:

$$(C_2H_2)e^{k_3 t} = k_2(CH_4)_0 \frac{k_1}{k_2 - k_1} \int (e^{-k_1 t} - e^{-k_2 t})e^{k_3 t}\, dt + \text{const}$$

$$= k_1 k_2 \frac{(CH_4)_0}{k_2 - k_1} \left[\frac{e^{-(k_1 - k_3)t}}{(k_3 - k_1)} - \frac{e^{-(k_2 - k_3)t}}{(k_3 - k_2)} \right] + \text{const.}$$

(6.8)

Now for the condition at $t = 0$ we obtain

$$\text{const} = \frac{-k_1 k_2}{k_2 - k_1}(CH_4)_0 \left\{ \frac{1}{(k_3 - k_1)} - \frac{1}{(k_3 - k_2)} \right\}.$$

(6.9)

Hence, by substitution in (6.8) and rearrangement, the resultant expression is

$$(C_2H_2) = \frac{k_1 k_2 (CH_4)_0}{k_2 - k_1} \left\{ \frac{e^{-k_1 t}}{(k_3 - k_1)} - \frac{e^{-k_2 t}}{(k_3 - k_2)} - e^{-k_3 t} \left[\frac{1}{k_3 - k_1} - \frac{1}{k_3 - k_2} \right] \right\}$$

$$= k_1 k_2 (CH_4)_0 \left\{ \frac{e^{-k_1 t}}{(k_2 - k_1)(k_3 - k_1)} + \frac{e^{-k_2 t}}{(k_3 - k_2)(k_1 - k_2)} + \frac{e^{-k_3 t}}{(k_3 - k_1)(k_3 - k_2)} \right\}.$$

(6.10)

Thus if values of the rate constants k_1, k_2 and k_3 are known as a function of temperature, the concentrations of ethylene and acetylene can be obtained for any given time spent in the reactor.

The concentration of ethylene and acetylene, relative to the initial concentration of methane, plotted as a function of time in the reactor is shown in Fig. 6.2, the reaction being considered to take place isothermally at 2000°K. It can be seen that both ethylene and acetylene concentrations pass through a maximum. Thus for a tubular or batch reactor there will be an optimum time leading to maximum possible acetylene yield.

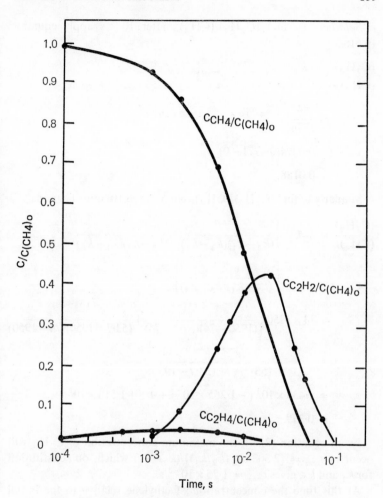

FIG. 6.2. Concentrations as a function of time in acetylene synthesis from methane (2003°K)

Example 6.1. Find the concentration of ethylene and acetylene relative to the initial methane concentration after 10^{-2} s in a tubular reactor if the operating temperature is 2000°K.

Calculate the time at which the maximum yield of ethylene occurs and its concentration relative to initial concentration of methane after that time.

$$k_1 = 74 \text{ s}^{-1} \qquad k_2 = 1950 \text{ s}^{-1} \qquad k_3 = 53 \cdot 7 \text{ s}^{-1}$$

Solution. To find $(C_2H_4)/(CH_4)_0$ after 10^{-2} s apply equation (6.6):

$$\frac{(C_2H_4)}{(CH_4)_0} = \frac{k_1}{k_2 - k_1} \{e^{-k_1 t} - e^{-k_2 t}\} \tag{6.6}$$

$$= \frac{74}{1876}\{e^{-74 \times 10^{-2}} - e^{-1950 \times 10^{-2}}\}$$

$$= 0 \cdot 0394(0 \cdot 4771 - 0)$$

$$= 0 \cdot 0188.$$

Similarly to find $(C_2H_2)/(CH_4)_0$ equation (6.10) must be used:

$$\frac{(C_2H_2)}{(CH_4)_0} = k_1 k_2 \left\{ \frac{e^{-k_1 t}}{(k_2 - k_1)(k_3 - k_1)} + \frac{e^{-k_2 t}}{(k_3 - k_2)(k_1 - k_2)} \right.$$

$$\left. + \frac{e^{-k_3 t}}{(k_3 - k_1)(k_3 - k_2)} \right\}$$

$$= 74 \times 1950 \left\{ \frac{e^{-74 \times 10^{-2}}}{(1950 - 74)(53 \cdot 7 - 74)} + \frac{e^{-1950 \times 10^{-2}}}{(53 \cdot 7 - 1950)(74 - 1950)} \right.$$

$$\left. + \frac{e^{-53 \cdot 7 \times 10^{-2}}}{(53 \cdot 7 - 74)(53 \cdot 7 - 1950)} \right\}$$

$$= 1 \cdot 433 \times 10^5 \, (-1 \cdot 255 \times 10^{-5} + 0 \quad + 1 \cdot 518 \times 10^{-5})$$

$$= 0 \cdot 379.$$

From equation (2.30), the maximum concentration of C_2H_4 will occur at $t_{max} = \{2 \cdot 303 \log (k_2/k_1)\}/(k_2 - k_1)$ which, on substitution for k_1 and k_2, gives $t_{max} = 1 \cdot 75 \times 10^{-3}$ s.

At this time the concentration of ethylene relative to the initial concentration of methane is given by equation (2.31):

$$\frac{(C_2H_4)_{max}}{(CH_4)_0} = \left(\frac{k_2}{k_1} \right)^{k_2/(k_1 - k_2)}$$

$$= \left(\frac{1950}{74} \right)^{1950/(74 - 1950)}$$

$$= 0 \cdot 033.$$

The calculated points are shown in Fig. 6.2.

6.2.3   PROCESSES FOR ACETYLENE FORMATION FROM HYDROCARBONS

From the analysis of the thermodynamics and kinetics of acetylene producing processes from hydrocarbons it has been deduced that the necessary reaction conditions are (a) high temperature ($>1500°$K for methane and $>1200°$K for naphtha) (b) a time of stay in the reactor much less than that necessary to bring about equilibrium. A quench is obviously a vital necessity to 'freeze' the acetylene once it is formed to prevent further dissociation.

Three different types of process have been developed to provide these conditions (a) partial oxidation (b) electric arc (c) thermal pyrolysis. The distinctive features of each and the form of the reactors used will be described in turn.

(a) Partial Oxidation Processes

Most acetylene from hydrocarbon sources is made by this process. There are many plants operating throughout the world producing acetylene in amounts greater than 50,000 tons p.a.

In this process, the heat required to achieve the necessary temperature of operation is furnished by the oxidation of part of the feed with pure oxygen. The oxidation reaction may be described by the equation

$$CH_4 + O_2 = CO + H_2O + H_2$$

where the hydrocarbon in this case is methane.

At $1500°$K this reaction is exothermic to the extent of $273 \cdot 2$ kJ. It is to be noted that in this methane rich reaction the combustion product is carbon monoxide rather than the dioxide. Pure oxygen is used rather than air to reduce the amount of gas to be handled and to eliminate any wastage of heat associated with heating the inert nitrogen gas present in air. However the use of pure oxygen does add to the cost since prior separation from the nitrogen is required.

The exothermic partial oxidation reaction thus generates the heat required in the endothermic pyrolysis of the feed to acetylene. In processes using natural gas, the feed and oxygen, in the molar ratio $1/0 \cdot 6$, are preheated separately to 870–$970°$K and are then ignited in a burner. Many of the processes in operation at the present time utilise the B.A.S.F. type reactor and a sketch of this is shown in Fig. 6.3. The main features of a burner are the mixing, burning and

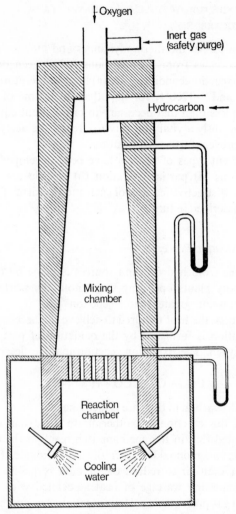

FIG. 6.3. Partial oxidation type reactor (B.A.S.F.) (Reprinted with permission from *Chem. Proc. Eng.*, October 1963, p. 579)

quench sections. Complete mixing of the feed streams must be attained. The mixture then enters the burning section through a burner block having a large number of small ports. A burner for 0·467 m³/s of methane feed has a burning zone of 38 mm diameter with a flame thickness of about 150 mm. The product from the burner zone is cooled with a water or oil spray to a temperature of about 310°K. The spray also helps to scrub out carbon from the effluent gases.

The operating pressure is slightly above atmospheric and the residence time is in the range 0·001–0·01 s. The composition of a typical effluent gas is given in Table 6.2; acetylene is present to the extent of about 8%.

TABLE 6.2

Composition of exit gases from partial oxidation burner

	% Volume		% Volume
Acetylene	8·0	Methane	5·0
Carbon dioxide	3·5	Hydrogen	56·0
Carbon monoxide	26·0	Oxygen	0·1
		Higher acetylenes	1·0

The kinetics of a partial oxidation process can be considered assuming that the reaction occurs in two stages, *viz*. partial combustion followed by pyrolysis of methane. The initial flame temperature can be calculated knowing the amount of heat evolved in the methane oxidation reaction, no pyrolysis being assumed to take place. Then, by assuming adiabatic conditions of operation, conversion to ethylene and acetylene can be calculated by means of equations (6.6) and (6.10) over a series of small time intervals, within each of which the reaction may be considered to occur isothermally. Thus in contrast to the previous case considered, the temperature over the complete time of reaction will not be isothermal but will be decreasing. A typical product distribution as a function of time and temperature is shown in Fig. 6.4. Because of simplifications made in the kinetics, and the assumption that the pyrolysis reaction follows the partial combustion reaction, there is obviously some degree of uncertainty in the results but the qualitative picture is satisfactory. The calculated product yields shown agree well with those obtained experimentally by sampling techniques, indicating that the model adopted, and rate constant data used, satisfactorily describe the observed behaviour.

As seen in Table 6.2 the partial combustion process produces large volumes of hydrogen and carbon monoxide, and commerical success

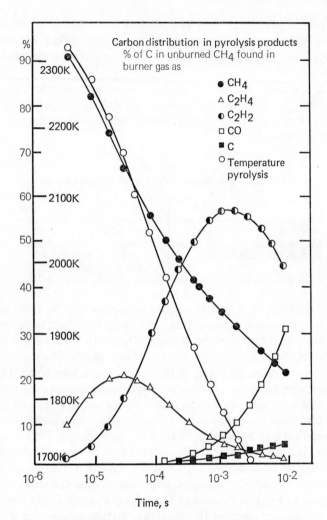

FIG. 6.4. Product distribution in partial oxidation process as a function of reactor residence time (Reprinted with permission from *Chem. Eng. Prog.* **57**, (11), 54 (1961))

depends to a large extent on the economic utilisation of these by-products. Separation and purification costs are also high and again economic operation can only be obtained if high acetylene outputs are achieved.

(b) Electric Arc Process

The electric arc process, commonly known as the Hüls process, for acetylene manufacture from hydrocarbon was the first to be developed. The earliest plants operated on a low boiling point

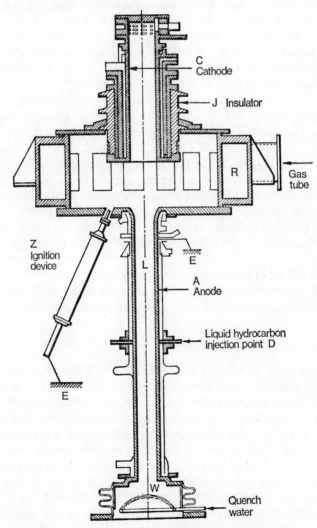

FIG. 6.5. Electric arc reactor (Hüls) (Reprinted with permission from *Chem. Proc. Eng.*, October 1963, p. 579)

petroleum fraction as the feedstock, but development since has led to processes which utilise either methane or ethane, l.p.g. or naphtha feed.

The form of the reactor used today is shown in Fig. 6.5. The d.c. electric arc is a means whereby a high temperature may be generated. The two water-cooled electrodes are made from iron, the cathode being bell shaped and the anode tubular in nature, of diameter 85–105 mm. The gas enters through a twisting guide chamber, 0·7–0·8 m in diameter and 0·3–0·5 m high, and passes down through the tubular anode where the arc is formed. The arc is about 1 m long and extends 0·4–0·5 m into the anode chamber.

The gases are then quenched with light hydrocarbon which recovers some of the heat, and finally by water to a temperature of 450°K. The secret of success in operation is to create a stable arc and it is for this reason that the gas is made to swirl, the arc burning in the centre of this swirl. The life of the electrode is about 800 h for the cathode and 150 h for the anode. Hence, to reduce stoppages for replacement, two furnaces are usually used in parallel. The peak reaction temperature in the arc is estimated to be about 1900°K.

In a typical process, for each 100 kg of acetylene produced, 290 kg of hydrocarbon, 1310 kWh electric energy and 150 kg of steam are required, while 49·5 kg ethylene, 29 kg carbon black, 17 kg residual oil and 280 m³ of hydrogen are also produced.

The greatest disadvantage of the arc process lies in the cost of electricity. It is supposed that a.c. operation would be cheaper.

(c) Thermal Pyrolysis Process

This process, commonly known as the Wulff process, is a cyclic one, in which the heat developed in previously heated corundum bricks, placed on either side of a combustion section, is transferred to the hydrocarbon to be decomposed. A diagram of the reactor is shown in Fig. 6.6. Air enters from the L.H. side and is preheated by passing through channels in the hot corundum bricks where it mixes with the fuel which burns. The mixture then flows through the channels within the corundum bricks to the R.H. side raising the temperature of the bricks to 1700–1800°K at the R.H. port. This part of the process lasts for one minute, after which feed plus steam enters the furnace from the R.H. side, passes between the now hot corundum and reaches operating temperature in the centre. During

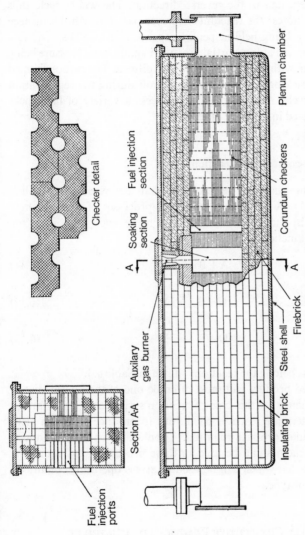

Checker detail

Soaking section

Fuel injection section

A

Auxiliary gas burner

A

Section A-A

Fuel injection ports

Plenum chamber

Corundum checkers

Firebrick

Steel shell

Insulating brick

Fig. 6.6. Thermal pyrolysis reactor (Wulff) (Reprinted with permission from *Chem. Proc. Eng.*, October, 1963, p. 579)

reaction the gas cools and leaves the reactor on the L.H. side. This part of the cycle operates for another minute, the first two stages then being repeated in the reverse direction. The whole cycle thus takes four minutes; the reactants however pass through the furnace in about 0·1 s, being held at peak temperature for 0·03 s. The furnaces are operated at low partial pressure in pairs, total pressure being about 0·5 atm. with steam being used as a diluent. The carbon formed during the cracking operation is burned off during the regeneration part of the cycle. As with the Hüls process a variety of feedstocks can be processed in this type of reactor.

6.3   Parallel-consecutive Reactions

An important class of reaction is that described as parallel-consecutive. Many prominent industrial processes of this type may be described by the following reaction scheme, each step displaying second order kinetics:

$$A + B \xrightarrow{k_1} C \tag{6.11}$$

$$C + B \xrightarrow{k_2} D \tag{6.12}$$

$$D + B \xrightarrow{k_3} E. \tag{6.13}$$

To such a class belong the reactions shown in Table 6.3.

The required product will most often be one of the intermediates, C or D, and hence for this type of reaction system, it is important to be able to determine reaction conditions to produce optimum yield of the particular product required. Tubular, batch or continuous stirred tank reactors can be used in these processes, and equations will be formulated which can be used to assess operating conditions for each reactor type.

6.3.1   PARALLEL-CONSECUTIVE REACTIONS IN A BATCH OR TUBULAR REACTOR

Consider reactions (6.11)–(6.13) taking place in a constant volume batch reactor. The following differential rate equations can be

TABLE 6.3

Examples of parallel-consecutive reactions of industrial importance

A	B	C	D	E
Water	Ethylene oxide	Ethylene glycol	Diethylene glycol	Triethylene glycol
Ammonia	Ethylene oxide	Monoethanolamine	Diethanolamine	Triethanolamine
Benzene	Chlorine	Monochlorobenzene	Dichlorobenzene	Trichlorobenzene
Methane	Chlorine	Methylchloride	Dichloromethane	Trichloromethane

written to describe the rate of change of concentration of each species:

$$\frac{dC_A}{dt} = -k_1 C_A C_B \tag{6.14}$$

$$\frac{dC_C}{dt} = k_1 C_A C_B - k_2 C_C C_B \tag{6.15}$$

$$\frac{dC_D}{dt} = k_2 C_C C_B - k_3 C_D C_B \tag{6.16}$$

$$\frac{dC_E}{dt} = k_3 C_D C_B. \tag{6.17}$$

The equations for a tubular reactor in which the density remains constant will be similar if t now refers to the time taken by the reactants to flow through a given section of the reactor.

In principle these four equations, together with a mass balance, can be solved to predict the concentration of each species as a function of time provided the individual rate constants are known. However a more limited objective, which requires a knowledge of the ratios of rate constants rather than their absolute values, involves an assessment of the degree of conversion of reactants A or B resulting in maximum yields of the required product, C or D. In this approach the time variable can be eliminated by taking ratios of concentrations, and all concentrations are expressed as a ratio with respect to the initial concentration of reactant A.

Elimination of the time variable results in first order linear differential equations which may be readily solved for the case when no products are present in the reactor feed. Equations relating C_C/C_{A_0} and C_D/C_{A_0} to C_A/C_{A_0} are deduced and C_E/C_{A_0} is obtained by a mass balance.

(a) Division of equation (6.15) by (6.14) and rearrangement leads to a first order linear differential equation

$$\frac{dC_C}{dC_A} - \alpha_2 \frac{C_C}{C_A} = -1 \tag{6.18}$$

where $\alpha_2 = k_2/k_1 \neq 1$.

Solution of this equation is obtained by multiplying by the factor

$$\int \exp\left[\frac{-\alpha_2}{C_A} dC_A\right] = \exp[-\alpha_2 \ln C_A] = \exp\left[\ln 1/C_A^{\alpha_2}\right] = \frac{1}{C_A^{\alpha_2}}$$

to give

$$C_C \frac{1}{C_A{}^{\alpha_2}} = \int -\frac{dC_A}{C_A{}^{\alpha_2}} + p$$

$$= \frac{C_A{}^{-\alpha_2+1}}{(\alpha_2-1)} + p \qquad (6.19)$$

where p is the constant of integration.

p is evaluated by applying the condition that at $C_A = C_{A_0}$, $C_C = 0$ which leads to a value of $-1/(\alpha_2-1)C_{A_0}{}^{\alpha_2-1}$.

Introducing this constant into equation (6.19) and rearranging gives

$$C_C = \frac{C_A{}^{\alpha_2}}{(\alpha_2-1)C_A{}^{\alpha_2-1}} - \frac{C_A{}^{\alpha_2}}{(\alpha_2-1)C_{A_0}{}^{\alpha_2-1}}$$

$$= \frac{C_A}{(\alpha_2-1)} - \frac{C_A{}^{\alpha_2}}{(\alpha_2-1)C_{A_0}{}^{\alpha_2-1}}$$

$$\text{or } \frac{C_C}{C_{A_0}} = \frac{1}{\alpha_2-1}(\gamma - \gamma^{\alpha_2}) \qquad (6.20)$$

where $\gamma = C_A/C_{A_0}$.

A special case exists when $k_1 = k_2$. In this case the differential equation which applies is

$$\frac{dC_C}{dC_A} - \frac{C_C}{C_A} = -1. \qquad (6.21)$$

This may be solved similarly, the integrating factor being $1/C_A$ and the constant of integration $\ln C_{A_0}$.

The solution to equation (6.21) is

$$\frac{C_C}{C_{A_0}} = -\gamma \ln \gamma. \qquad (6.22)$$

(b) The differential equation from which C_D may be obtained is derived by dividing equation (6.16) by (6.14) and substituting for C_C:

$$\frac{dC_D}{dC_A} - \alpha_3 \frac{C_D}{C_A} = \frac{-\alpha_2}{(\alpha_2-1)}\left\{1 - \left(\frac{C_A}{C_{A_0}}\right)^{\alpha_2-1}\right\} \qquad (6.23)$$

where $\alpha_3 = k_3/k_1$.

The integrating factor here is $1/C_A{}^{\alpha_3}$ and the solution to equation (6.23) becomes

$$C_D \frac{1}{C_A{}^{\alpha_3}} = \int \frac{-\alpha_2}{(\alpha_2 - 1)} \left\{ 1 - \left(\frac{C_A}{C_{A_0}}\right)^{\alpha_2 - 1} \right\} \frac{dC_A}{C_A{}^{\alpha_3}} + p.$$

$$= \frac{-\alpha_2}{(\alpha_2 - 1)} \left[\frac{C_A{}^{1-\alpha_3}}{(1-\alpha_3)} - \frac{1}{C_{A_0}{}^{\alpha_2 - 1}} \frac{C_A{}^{\alpha_2 - \alpha_3}}{(\alpha_2 - \alpha_3)} \right] + p. \qquad (6.24)$$

p is evaluated by applying the condition that at $C_A = C_{A_0}$, $C_D = 0$ to give

$$p = \frac{\alpha_2}{(1-\alpha_3)} \frac{C_{A_0}{}^{1-\alpha_3}}{(\alpha_2 - \alpha_3)}.$$

Substitution for p in (6.24) and rearrangement gives

$$\frac{C_D}{C_{A_0}} = \alpha_2 \left[\frac{\gamma}{(1-\alpha_2)(1-\alpha_3)} + \frac{\gamma^{\alpha_2}}{(\alpha_2 - 1)(\alpha_2 - \alpha_3)} + \frac{\gamma^{\alpha_3}}{(1-\alpha_3)(\alpha_2 - \alpha_3)} \right].$$

$$(6.25)$$

(c) C_E/C_{A_0} is then obtained from the mass balance

$$C_{A_0} = C_A + C_C + C_D + C_E. \qquad (6.26)$$

Substitution for C_C and C_D using equations (6.20) and (6.25) leads to

$$\frac{C_E}{C_{A_0}} = 1 - \gamma - \frac{1}{(\alpha_2 - 1)} \left\{ \gamma - \gamma^{\alpha_2} \right\} - \alpha_2 \left\{ \frac{\gamma}{(1-\alpha_2)(1-\alpha_3)} \right.$$

$$\left. + \frac{\gamma^{\alpha_2}}{(\alpha_2 - 1)(\alpha_2 - \alpha_3)} + \frac{\gamma^{\alpha_3}}{(1-\alpha_3)(\alpha_2 - \alpha_3)} \right\}$$

$$= 1 - \frac{\gamma \alpha_2 \alpha_3}{(1-\alpha_2)(1-\alpha_3)} - \frac{\gamma^{\alpha_2} \alpha_3}{(\alpha_2 - 1)(\alpha_2 - \alpha_3)} - \frac{\alpha_2 \gamma^{\alpha_3}}{(1-\alpha_3)(\alpha_2 - \alpha_3)}.$$

$$(6.27)$$

(d) It is worthy of note that the concentration of C relative to the initial concentration of A at its maximum value, and the conversion of A at which this maximum is found, may be obtained by finding dC_C/dC_A and setting equal to zero.

From equation (6.20)

$$\beta = \frac{C_C}{C_{A_0}} = \frac{1}{\alpha_2 - 1} \left(\gamma - \gamma^{\alpha_2} \right).$$

On differentiating and equating to zero we obtain the value of C_A/C_{A_0} at which C_C/C_{A_0} has its maximum value:

$$\frac{d\beta}{d\gamma} = 1 - \alpha_2\,\gamma^{\alpha_2 - 1} = 0$$

or
$$\gamma_{max} = \left(\frac{1}{\alpha_2}\right)^{1/\alpha_2 - 1}. \tag{6.28}$$

The value of $(C_C/C_{A_0})_{max}$ is then obtained by substituting this value of γ_{max} for γ in equation (6.20)

$$\left(\frac{C_C}{C_{A_0}}\right)_{max} = \frac{1}{\alpha_2 - 1}\left\{\left(\frac{1}{\alpha_2}\right)^{1/\alpha_2 - 1} - \left(\frac{1}{\alpha_2}\right)^{\alpha_2/\alpha_2 - 1}\right\}. \tag{6.29}$$

(e) So far nothing has been said regarding the concentration of the other reactant B. This can be readily derived by means of a mass balance:

$$\frac{C_{B_0}}{C_{A_0}} = \frac{C_B}{C_{A_0}} + \frac{C_C}{C_{A_0}} + \frac{2C_D}{C_{A_0}} + \frac{3C_E}{C_{A_0}}. \tag{6.30}$$

It is seen that C_B/C_{A_0} can be obtained in terms of C_A/C_{A_0} from the expressions for C_C, C_D and C_E derived above.

6.3.2 PARALLEL-CONSECUTIVE REACTIONS IN A STIRRED TANK REACTOR

Consider the three-stage reaction system given by equations (6.11)–(6.13) taking place in a stirred tank reactor. For the stirred tank reactor, the concentrations of intermediates relative to that of A initially, can be obtained algebraically by carrying out mass balances for each species. No differential equations are thus necessary.

By a mass balance for A

$$C_{A_0} = C_A + \bar{t}(-r_A) = C_A + \bar{t}k_1 C_A C_B \tag{6.31}$$

where $\bar{t}$ is the mean holding time in the stirred tank reactor.

Similarly by a mass balance for C

$$C_{C_0} = C_C + \bar{t}(-r_C) = C_C - \bar{t}k_1 C_A C_B + \bar{t}k_2 C_C C_B. \tag{6.32}$$

Equation (6.31) is rearranged to provide an expression for C_B which is then substituted in equation (6.32) to give C_C/C_{A_0} in terms of C_A/C_{A_0}, it again being assumed the product concentrations are zero initially, i.e. $C_{C_0} = 0$:

$$C_C = \frac{C_A (C_{A_0} - C_A)}{C_A + k_2/k_1 (C_{A_0} - C_A)}$$

or

$$\frac{C_C}{C_{A_0}} = \frac{\gamma(1-\gamma)}{\gamma + \alpha_2(1-\gamma)}. \tag{6.33}$$

The equivalent expression for C_D/C_{A_0}, obtained by substituting equation (6.31) in the appropriate mass balance equation for D, is

$$\frac{C_D}{C_{A_0}} = \frac{\alpha_2(C_C/C_{A_0})(1-\gamma)}{\gamma + \alpha_3(1-\gamma)} \tag{6.34}$$

where $\alpha_3 = k_3/k_1$.

For an n step reaction the general expression is

$$\frac{C_i}{C_{A_0}} = \frac{\alpha_i(C_{i-1}/C_{A_0})(1-\gamma)}{\gamma + \alpha_{i+1}(1-\gamma)}. \tag{6.35}$$

For the three step reaction being considered the concentration of final product, E, is deduced from a mass balance for E

$$C_{E_0} = C_E + \bar{t}(-k_3 C_D C_B) = 0.$$

Substitution for C_B from equation (6.31) leads to

$$\frac{C_E}{C_{A_0}} = \alpha_3\left(\frac{C_D}{C_{A_0}}\right)\frac{(1-\gamma)}{\gamma}. \tag{6.36}$$

As in the case of a batch reactor $(C_C/C_{A_0})_{max}$ is obtained by setting $dC_C/dC_A = 0$.

From equation (6.33)

$$\beta = \frac{C_C}{C_{A_0}} = \frac{\gamma(1-\gamma)}{\gamma + \alpha_2(1-\gamma)},$$

so that

$$\frac{d\beta}{d\gamma} = \frac{(\gamma + \alpha_2 - \alpha_2\gamma)(1-2\gamma) - (\gamma - \gamma^2)(1-\alpha_2)}{\{\gamma + \alpha_2(1-\gamma)\}^2} = 0,$$

or

$$(\alpha_2 - 1)\gamma^2 - 2\alpha_2\gamma + \alpha_2 = 0. \tag{6.37}$$

Solution of equation (6.37) gives

$$\gamma_{max} = \frac{2\alpha_2 \pm [4\alpha_2^2 - 4\alpha_2(\alpha_2 - 1)]^{\frac{1}{2}}}{2(\alpha_2 - 1)}. \tag{6.38}$$

$(C_C/C_{A_0})_{max}$ is then obtained by substituting this value of γ_{max} for γ in equation (6.33).

6.3.3 HYDROCARBON CHLORINATION REACTION

The chlorination of methane leading to the series of products methyl chloride, methylene dichloride, chloroform and carbon tetrachloride, and the chlorination of benzene to give chlorobenzene, di-, tri- and higher substituted benzenes, are industrial processes which follow the pattern of reaction outlined above.

By proper choice of reaction conditions it should be possible to obtain substantial yields of the substituted compound required. An important factor influencing the relative concentrations of the possible products will be the magnitude of the rate constant ratios, α_i. These will be fixed at a given temperature, but if the rate constants of the individual reactions are affected by temperature to a different extent, i.e. have different values of activation energy, then some control can be maintained over relative yields by judicious choice of operating temperature.

A much more effective method of controlling the product distribution is by correct choice of reactant ratio. It would be expected that as the chlorine to hydrocarbon ratio increases, the extent of successive chlorination would increase. The significance of reactant ratio will be demonstrated quantitatively for the chlorination of benzene.

Consider the first three steps in benzene chlorination:

$$\text{Benzene} + \text{Cl}_2 \xrightarrow{k_1} \text{chlorobenzene} + \text{HCl}$$

$$\text{Chlorobenzene} + \text{Cl}_2 \xrightarrow{k_2} \text{dichlorobenzene} + \text{HCl}$$

$$\text{Dichlorobenzene} + \text{Cl}_2 \xrightarrow{k_3} \text{trichlorobenzene} + \text{HCl}.$$

Suppose that the reaction is to be carried out (a) batchwise (b) in a CSTR. It is actually performed in a semi-batchwise fashion with chlorine being bubbled continuously into the reactor, but if the assumption is made that chlorine is added slowly enough so that its concentration and that of the HCl produced are small, and the density and volume of the solution remain constant, then the expressions applicable for a constant volume batch and CST reactors stated in sections 6.3.1 and 6.3.2 will be valid.

The rate expressions are those given by equations (6.14–6.17) where $A \equiv$ benzene, $B \equiv$ chlorine, $C \equiv$ chlorobenzene, $D \equiv$ dichlorobenzene and $E \equiv$ trichlorobenzene.

Equations (6.20), (6.26) and (6.27) are used to calculate the concentration ratios chlorobenzene/(benzene)$_{initial}$, dichlorobenzene/(benzene)$_{initial}$ and trichlorobenzene/(benzene)$_{initial}$ respectively for the batch case, and (6.33), (6.34) and (6.36) the same ratios for the CSTR. These concentration ratios may be determined either as a function of benzene remaining unreacted or in terms of the moles chlorine which have been used/mole of benzene initially present.

Example 6.2. Plot the values of the composition of a mixture of chlorinated benzenes up to and including the tri-substituted compound against moles of chlorine consumed per mole of benzene charged for reaction taking place in (a) a constant volume batch reactor (b) CSTR at 328°K.

McMullin (1948) quotes value of $k_2/k_1 = 0.125$ and $k_3/k_1 = 0.0042$ at this temperature.

Carry out the calculation up to moles chlorine consumed/mole benzene charged of 2.2.

Solution. Batch. To serve as a representative calculation consider the mixture composition when benzene is 90% consumed, i.e. $\gamma = C_A/C_{A_0} = 0.1$. The concentration of chlorobenzene, dichlorobenzene and trichlorobenzene relative to that of benzene initially is obtained by application of equations (6.20), (6.25) and (6.27).

In this example $\alpha_2 = 0.125$ and $\alpha_3 = 0.0042$.

$$\frac{C_C}{C_{A_0}} = \frac{1}{\alpha_2 - 1}(\gamma - \gamma^{\alpha_2})$$

$$= \frac{1}{0.125 - 1}(0.1 - 0.1^{0.125})$$

$$= 0.744.$$

$$\frac{C_D}{C_{A_0}} = \alpha_2\left[\frac{\gamma}{(1-\alpha_2)(1-\alpha_3)} + \frac{\gamma^{\alpha_2}}{(\alpha_2-1)(\alpha_2-\alpha_3)} + \frac{\gamma^{\alpha_3}}{(1-\alpha_3)(\alpha_2-\alpha_3)}\right]$$

$$= 0.125\left[\frac{0.1}{(0.875)(0.9958)} + \frac{0.1^{0.125}}{(-0.875)(0.1208)} + \frac{0.1^{0.0042}}{(0.9958)(0.1208)}\right]$$

$$= 0.152.$$

$$\frac{C_E}{C_{A_0}} = 1 - \frac{\gamma\alpha_2\alpha_3}{(1-\alpha_2)(1-\alpha_3)} - \frac{\gamma^{\alpha_2}\alpha_3}{(\alpha_2-1)(\alpha_2-\alpha_3)} - \frac{\alpha_2\gamma^{\alpha_3}}{(1-\alpha_3)(\alpha_2-\alpha_3)}$$

$$= 1 - \frac{(0\cdot1)(0\cdot125)(0\cdot0042)}{(0\cdot875)(0\cdot9958)} - \frac{(0\cdot1^{0\cdot125})(0\cdot0042)}{(-0\cdot875)(0\cdot1208)}$$

$$- \frac{(0\cdot125)(0\cdot1^{0\cdot0042})}{(0\cdot9958)(0\cdot1208)}$$

$$= 0\cdot004.$$

The moles of chlorine reacted/mole of initial benzene is given by equation (6.30):

$$\frac{C_{B_0}-C_B}{C_{A_0}} = \frac{C_C}{C_{A_0}} + \frac{2C_D}{C_{A_0}} + \frac{3C_E}{C_{A_0}}$$

$$= 0\cdot744 + 0\cdot304 + 0\cdot012$$

$$= 1\cdot060.$$

CSTR. In this case C_C/C_{A_0}, C_D/C_{A_0} and C_E/C_{A_0} are determined from equations (6.33), (6.34) and (6.36):

$$\frac{C_C}{C_{A_0}} = \frac{\gamma(1-\gamma)}{\gamma+\alpha_2(1-\gamma)}$$

$$= \frac{(0\cdot1)(0\cdot9)}{(0\cdot1)+(0\cdot125)(0\cdot9)} = 0\cdot424.$$

$$\frac{C_D}{C_{A_0}} = \frac{\alpha_2\left(\dfrac{C_C}{C_{A_0}}\right)(1-\gamma)}{\gamma+\alpha_3(1-\gamma)}$$

$$= \frac{(0\cdot125)(0\cdot424)(0\cdot9)}{0\cdot1+(0\cdot0042)(0\cdot9)} = 0\cdot459.$$

$$\frac{C_E}{C_{A_0}} = \alpha_3\left(\frac{C_D}{C_{A_0}}\right)\left(\frac{1-\gamma}{\gamma}\right)$$

$$= (0\cdot0042)(0\cdot459)\frac{(0\cdot9)}{(0\cdot1)} = 0\cdot017.$$

Again the moles of chlorine consumed/mole of benzene added

initially is given by

$$\frac{C_{B_0} - C_B}{C_{A_0}} = \frac{C_C}{C_{A_0}} + \frac{2C_D}{C_{A_0}} + \frac{3C_E}{C_{A_0}}$$

$$= 0 \cdot 424 + 0 \cdot 918 + 0 \cdot 051$$

$$= 1 \cdot 393.$$

Complete results for the two cases are shown in Tables 6.4 and 6.5 and graphically in Fig. 6.7.

TABLE 6.4

Batch reactor. Product composition in mixed chlorobenzenes based on 1 mole benzene charged.

Benzene remaining $\equiv C_A/C_{A_0}$	Monochloro-benzene $\equiv C_C/C_{A_0}$	Dichloro-benzene $\equiv C_D/C_{A_0}$	Trichloro-benzene $\equiv C_E/C_{A_0}$	Moles chlorine reacted per mole benzene charged $\equiv (C_{B_0} - C_B)/C_{A_0}$
0·8	0·197	0·003	0·0	0·203
0·5	0·477	0·021	0·002	0·525
0·2	0·706	0·091	0·003	0·897
0·1	0·744	0·152	0·004	1·060
0·01	0·632	0·352	0·006	1·354
10^{-3}	0·482	0·509	0·008	1·524
10^{-6}	0·203	0·765	0·032	1·829
10^{-10}	0·064	0·877	0·059	1·995
10^{-20}	0·004	0·852	0·144	2·140
10^{-30}	0·0	0·773	0·227	2·227

TABLE 6.5

CSTR. Product composition in mixed chlorobenzenes based on 1 mole benzene charged

Benzene remaining $\equiv C_A/C_{A_0}$	Monochloro-benzene $\equiv C_C/C_{A_0}$	Dichloro-benzene $\equiv C_D/C_{A_0}$	Trichloro-benzene $\equiv C_E/C_{A_0}$	Moles chlorine reacted per mole benzene charged $\equiv (C_{B_0} - C_B)/C_{A_0}$
0·8	0·194	0·006	0·0	0·206
0·5	0·4445	0·0555	0·0	0·5555
0·3	0·542	0·157	0·001	0·859
0·1	0·424	0·459	0·017	1·393
0·05	0·282	0·619	0·049	1·667
0·03	0·192	0·685	0·093	1·841
0·01	0·074	0·647	0·269	2·175

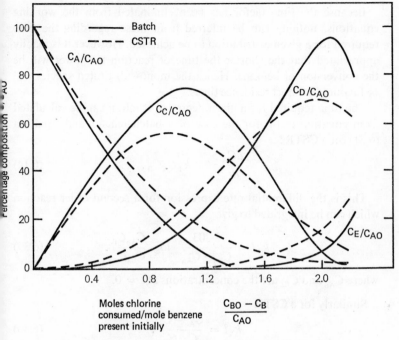

FIG. 6.7. Distribution of products in benzene chlorination

The following points should be noted:

(1) At the same level of chlorine consumed the extent of benzene decomposition is always greater for the batch case than for the CST case.

(2) For both the mono and dichlorobenzenes, maximum concentrations are greater in the case of the batch reactor than for the CST case and selectivities at these two points are likewise better for the batch than the CST case, i.e. C_C/C_D and C_D/C_E are higher in the batch reactor than in a continuous stirred tank reactor.

(3) For a high concentration of chlorobenzene relative to the higher substituted compounds, the moles of chlorine used must be kept small, while the opposite is the case if higher substituted compounds are required. These two objectives may be achieved by maintaining an excess of benzene over chlorine for the first and vice-versa for the second.

Because the time factor has been eliminated from the working equations, nothing can be inferred from these regarding the time required for a given distribution to be achieved. However it is readily appreciated that the shorter the time of reaction the lower will be the conversion of benzene. Hence the mono-substituted species will be favoured at short residence times.

The time required for a given degree of reaction can be calculated from equation (6.14) for the case of a batch reactor and equation (6.31) for a CSTR:

$$\frac{dC_A}{dt} = -k_1 C_A C_B. \tag{6.14}$$

This is the differential rate expression for a second order reaction which can be integrated to give

$$k_1 t = \frac{2 \cdot 303}{C_{B_0} - C_{A_0}} \log \frac{C_{A_0} C_B}{C_{B_0} C_A}, \tag{6.39}$$

where C_{A_0} and C_{B_0} are the concentrations at $t = 0$.

Similarly for a CSTR

$$k_1 \bar{t} = \frac{C_{A_0} - C_A}{C_A \, C_B}. \tag{6.40}$$

The effect of increasing the ratio C_{B_0}/C_{A_0} can be deduced by considering equation (6.39). For a given conversion, or value of C_A/C_{A_0}, the moles of chlorine consumed/mole of benzene charged, $(C_{B_0} - C_B)/C_{A_0}$, is fixed. Hence C_B increases as C_{B_0} increases, but for an increase in C_{B_0}/C_{A_0}, $C_{B_0} - C_{A_0}$ changes to a greater extent than the logarithmic term. Thus as C_{B_0}/C_{A_0} increases, the time for a given conversion decreases or, alternatively, for the same time spent in the reactor the conversion will increase, i.e. C_A/C_{A_0} will decrease, and consequently more of the higher substituted compounds will be formed.

Industrially the chlorination of benzene is carried out either batchwise or in continuous stirred tanks. In the batch case the temperature is maintained below 318°K. Ferric chloride is added as a catalyst (1 kg per 700 kg of mixture). In the continuous case small externally cooled steel vessels are used, chlorine being added to each one. With large benzene to chlorine ratios at temperatures between 293–313°K, selectivities of 95% to chlorobenzene have been reported (Kirk-Othmer, 1963).

REFERENCES

CURTIS, E. H. 1963. *Chem. Prog. Eng.* **44,** 579.

KIRK-OTHMER. 1963. *Encyclopaedia of chemical technology*, 2nd Ed. Interscience, New York, Vol. 5, p. 257.

LEROUX, P. J. and MATHIEU, P. M. 1961. *Chem. Eng. Prog.* **57,** (11), 54.

MCMULLIN, R. B. 1948. *Chem. Eng. Prog.* **44,** 183.

7

EXOTHERMIC CATALYTIC REACTIONS—PHTHALIC ANHYDRIDE SYNTHESIS

7.1 Introduction

There are a number of highly exothermic gas phase reactions being carried out in the presence of solid materials acting as catalysts. A selection of these important reactions is listed in Table 7.1.

TABLE 7.1

Some industrial exothermic catalytic reactions

		Reaction temperature, °K	$\Delta H°_{298°K}$ kJ/mol
1.	$C_6H_6 + 3H_2 \rightarrow C_6H_{12}$ benzene cyclohexane	420–530	$-206{\cdot}1$
2.	$C_8H_{10} + 3O_2 \rightarrow C_8H_4O_3 + 3H_2O$ o-xylene phthalic anhydride	610–720	$-1127{\cdot}9$
3.	$C_{10}H_8 + 4\frac{1}{2}O_2 \rightarrow C_8H_4O_3 + 2CO_2 + 2H_2O$ naphthalene phthalic anhydride	610–750	$-1753{\cdot}2$
4.	$C_2H_4 + \frac{1}{2}O_2 \rightarrow C_2H_4O$ ethylene ethylene oxide	510–560	$-103{\cdot}3$
5.	$C_4H_8 + 3O_2 \rightarrow C_4H_2O_3 + 3H_2O$ butylene maleic anhydride	700–750	$-1133{\cdot}9$
6.	$C_2H_2 + HCl \rightarrow C_2H_3Cl$ acetylene vinyl chloride	380–450	$-103{\cdot}1$

Typical temperatures of reaction and heat evolved per mole of key reactant at 298°K are also shown in Table 7.1. These heats of reaction have been calculated from the standard heats of formation of the relevant reactants and products.

For this class of reaction the problem of heat removal is obviously of paramount importance. It can be seen that oxidation of either o-xylene or naphthalene to phthalic anhydride is among the most exothermic of the reactions listed, and the latter will be used as the main example in the discussion in this chapter, although much that will be said regarding design and operation of the reactors can be applied generally.

7.2 Phthalic Anhydride Synthesis from Naphthalene

The production of phthalic anhydride by oxidation of naphthalene has long been of industrial significance because of the importance of the anhydride as an intermediate in the dyestuffs industry, and the demand for the anhydride has increased enormously in recent years because of its widespread use in the manufacture of plasticisers, alkyd resins and as a constituent of polyester materials. Because of the decreasing amount of naphthalene available from coal tar sources and because the amount of naphthalene from petroleum sources is limited, processes have more recently been introduced based on o-xylene as feedstock. In many respects there is strong similarity between the features of the two processes; for example, reactors have been developed which will operate on either feedstock.

7.2.1 THERMODYNAMICS OF THE PROCESS

The chemical reaction for the oxidation is given by

$$\text{(naphthalene)} + 4\tfrac{1}{2}O_2 \rightarrow \text{(phthalic anhydride)} + 2CO_2 + 2H_2O. \qquad (7.1)$$

Experimental data on the standard heats and free energies of formation of gaseous naphthalene and phthalic anhydride are not readily available, although values of ΔH_f° and ΔG_f° of -461 and -331 kJ/mol have been reported at $298^\circ K$ for solid phthalic anhydride, and $60 \cdot 3$ kJ for ΔH_f° for solid naphthalene at $298^\circ K$ (Green, 1961). Hence the gaseous standard heats of formation of the compounds, from which the heat of reaction shown in Table 7.1 has been computed, have been obtained using the group contribution estimation method of Franklin (in Reid and Sherwood, 1966). These values and others at $600^\circ K$, which is near the reaction temperature, together with values for ΔG_f° calculated by the method of van Krevelen (in Reid and Sherwood, 1966), are listed in Table 7.2 below.

TABLE 7.2

Standard heats and free energies of formation of reactants and products in naphthalene oxidation

| | 298°K | | 600°K | |
	$\Delta H_f°$ kJ/mol	$\Delta G_f°$ kJ/mol	$\Delta H_f°$ kJ/mol	$\Delta G_f°$ kJ/mol
Naphthalene (g)	146·1	207	129·4	283·2
Phthalic Anhydride (g)	−326·3	−163·7†	−335·5*	−106†
CO_2 (g)	−398·2	−395	−394	−395·2
H_2O (g)	−242·2	−228·5	−245	−214

* This value for phthalic anhydride is based on the contribution from the anhydride grouping at 298°K.

† The values for phthalic anhydride may be subject to considerable error because the van Krevelen method is unreliable for anhydrides.

Using the above values the following standard heats and free energies of reaction are calculated for reaction (7.1):

$$(\Delta H°)_{298°K} = -326·3 + 2(-398·2) + 2(-242·2) - 146·1$$
$$= -1753·2 \text{ kJ}$$
$$(\Delta H°)_{600°K} = -335·5 + 2(-394) + 2(-245) - 129·4$$
$$= -1742·9 \text{ kJ}$$
$$(\Delta G°)_{298°K} = -163·7 + 2(-395) + 2(-228·5) - 20·7$$
$$= -1617·7 \text{ kJ}$$
$$(\Delta G°)_{600°K} = -106 + 2(-395·2) + 2(-214) - 283·2$$
$$= -1607·6 \text{ kJ}.$$

It is obvious from the $\Delta G°$ values that the oxidation reaction may be considered substantially complete both at the base temperature of 298°K and at the typical reaction temperature of 600°K, i.e. when equilibrium is reached the product yield is effectively 100%.

It should be noted also from the above data that the standard heat of reaction does not appear to depend much on the temperature level. The reaction being discussed is of course extremely exothermic. The value of −1743 kJ for oxidation of 1 g.mole of naphthalene may be compared with those of −46 and −98 kJ for the synthesis at 298°K of 1 g.mole of ammonia and for the oxidation of 1 g.mole sulphur dioxide to trioxide respectively.

7.2.2 KINETICS OF THE PROCESS

The mechanism of phthalic anhydride formation from naphthalene is undoubtedly complex. By-products identified in the reaction include 1,4-naphthoquinone and maleic anhydride together with carbon oxides and water. Based on a product distribution including these compounds, De Maria *et al.* (1961) postulated the following reaction scheme:

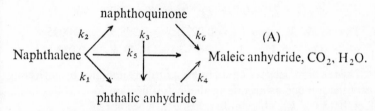

From studies of the kinetics of this series of reactions under carefully chosen conditions, a simplified reaction scheme for use in reactor design or for analysis of reactor performance was deduced in the following manner.

As the conversion of naphthalene tended to zero, the yields of phthalic anhydride and naphthoquinone extrapolated to near 50% of each with little gas formation. It may be concluded therefore that reaction (5) above is unimportant and that the rate constants for (1) and (2) can be taken equal.

Further, by examination of the products in naphthoquinone oxidation, little carbon dioxide or water formation was noted other than that which seemed to be formed through reactions (3) and (4). Hence reaction (6) may be considered to be very slow.

The simplified reaction scheme may thus be written

$$
NA \underset{k_1}{\overset{k_2}{<}} \quad \underset{PA}{\overset{NQ}{}} \Bigg| k_3 \quad \xrightarrow{k_4} MA, CO_2, H_2O \qquad (B)
$$

where NA = naphthalene, NQ = naphthoquinone, PA = phthalic

anhydride and MA = maleic anhydride. For each of the above reactions oxygen reacts with the organic compound, but since it is present in large excess (typically $99:1/O_2:NA$) its concentration will change but little during reaction, and pseudo-order rate constants, incorporating a factor for the constant oxygen concentration, can be derived for each step. To a first approximation each of the reactions may be considered first order with respect to the organic materials. Values of the rate constants for reaction on two catalysts were reported by de Maria; for catalyst A the following expressions may be used to calculate rate constant values at any temperature:

$$k_1 = k_2 = \exp(-1{\cdot}481 + 0{\cdot}055T) \quad k_3 = \exp(-0{\cdot}92 + 0{\cdot}0654T)$$
$$k_4 = \exp(-4{\cdot}747 + 0{\cdot}0268T) \quad \text{where } T = T°K - 593.$$

Thus at $593°K$, $k_1 = k_2 = 0{\cdot}227$, $k_3 = 0{\cdot}398$, $k_4 = 0{\cdot}0095 \text{ s}^{-1}$, while at $643°K$, $k_1 = k_2 = 3{\cdot}57$, $k_3 = 10{\cdot}49$, $k_4 = 0{\cdot}033 \text{ s}^{-1}$.

The reaction network could be even further simplified by neglecting naphthoquinone as an intermediate. Justification for this lies in the fact that $k_3 > k_2$. This simple scheme then becomes

$$\text{Naphthalene} \xrightarrow{k_1} \text{phthalic anhydride} \xrightarrow{k_4} CO_2, H_2O. \quad (C)$$

Values for the ratio k_1/k_4 at $593°K$ and $643°K$ of 50 and 200 have been quoted by Carberry and White (1969).

Because of the fact that phthalic anhydride can itself be oxidised further and, since according to the more complex model its formation can take place through two parallel routes, reaction time and temperature of operation, which can both affect the relative magnitude of the product concentrations, are obviously important variables in the optimisation of the process.

Details of existing processes for phthalic anhydride production from naphthalene will be given in the next section and the choice of operating conditions will be discussed in the light of the known kinetics and thermal effects in the reaction.

Before considering these aspects it should be pointed out that since the reaction involves more than just the simple step given by equation (7.1), the heat evolution/mole of naphthalene reacted will differ from that calculated in section 7.2.1. For example, the reaction

$$PA + 4\tfrac{1}{2}O_2 \rightarrow MA + 4CO_2 + H_2O$$

has a heat evolution of 1840 kJ at $298°K$. The overall heat of reaction will thus depend on the extent to which each of the reactions 1–4 takes place. The heat released per g.mole of naphthalene reacting has been estimated to be about 2135 kJ.

7.2.3 Operating Features of the Phthalic Anhydride Processes

The problem of heat removal has been seen to be of critical importance in this process. Operation using a fluidised bed of catalyst particles would seem at first sight to have considerable advantages with regard to the ease of heat removal over the alternative process which employs a fixed catalyst bed, since heat is much more readily dissipated in a continuously moving bed of hot particles than from a static solid mass in which heat is generated. In fact both types of process have been extensively used in recent years. Further subdivision can be made on the basis of catalyst type employed. 'American' processes employ a catalyst containing 10% V_2O_5 on alundun, silica or pumice while the 'German' process uses a catalyst consisting of 10% V_2O_5, $20–30\%$ K_2SO_4 on a silica gel support, the K_2SO_4 apparently poisoning the silica gel to some extent, since this support for the 'American' catalyst leads to complete oxidation to carbon dioxide and water.

As can be seen from the following table, which gives representative data for the various processes, the 'American' catalyst appears to be much more active than the 'German' since the required contact time in the fixed bed is shorter. On the other hand the resultant yield is markedly reduced. In fluidised bed processes a modified version of the 'German' type catalyst is universally used.

Despite the fact that temperature control would appear to be easier using a fluidised bed rather than a fixed bed, the bulk of anhydride is produced in fixed bed processes. As will be seen later, problems connected with temperature control in fixed beds can be, and have been, overcome. Another apparent attraction of the fluidised process, as can be noted from the table, is the smaller volumetric air requirement. The reason for this is that it is easier to maintain the mixture outside the explosive region in the fluidised than in the fixed bed process because of the intimacy of mixing of gas with the fine particles of fluidised catalyst, the actual reaction volume comprising a multitude of interstitial spaces of non-explosive dimensions. However the advantages of the fluidised bed are offset to a considerable degree by

TABLE 7.3

Comparison of operating conditions in reactors for phthalic anhydride production from naphthalene*

| | Fixed bed | | Fluidised bed | |
	American	German	American	German
Temp., °K	670–750	610–650	630	640
Pressure (atm.)	1·75	0·5	1	1
Contact Time (s)	0·6	4·2	19	19
kg air/kg $C_{10}H_8$	25–30	30	15	15
kg charge/s m^2 reactor cross section	0·053	0·022	0·024	0·024
kg charge/s m^3 reactor volume	0·09	0·009	0·0045	0·0045
kg catalyst/kg charge h	5·9	13·4	37·5	37·5
Yield (kg PA/100 kg NA) (max. 115)	80	104	80	100

* From Ruthruff, 1953.

the greater pressure drop resulting from the greater height of the dense bed, and resistances to flow resulting from the necessity of filters for the catalyst particles. Again larger yields might be expected because of the use of larger reactors in fluidised bed operation but these are reduced because of bypassing and channelling in the reactor.

Fixed bed reactors employ narrow tubes of about 20–50 mm diameter, molten salt (consisting of a mixture of $NaNO_3$ and KNO_3), which can be used to generate H.P. steam, being used as the heat transfer medium. In the BASF process (Spitz, 1968) a reactor containing 8927 tubes, each 42 mm in diameter and 3m long, has been employed to produce 14000 tons/year of anhydride. Complete conversion occurs for a feedstock containing 1 % naphthalene in air.

7.2.4 SOME FACTORS INFLUENCING THE CHOICE OF OPERATING CONDITIONS FOR FLUIDISED BED REACTORS

The greater ease of control of temperature in a fluidised bed reactor as compared to a fixed bed has already been noted. The possibility of operating near isothermal conditions therefore exists and, bearing in mind the relatively complex reaction mechanism involving parallel and series reactions in the anhydride production, it is possible that an optimum operating temperature could be chosen consistent with a given mean residence time in the reactor.

Equations can be derived for the concentration of each species present, relative to the initial naphthalene concentration, as a function of temperature and time spent in the reactor for the two extreme cases of plug flow and complete mixing in the reactor. It then remains to decide which of the models more closely fits the degree of mixing in a fluidised bed.

For a reactor considered to be of the tubular plug flow type, the concentrations can be derived by integration of the appropriate equations for the rate of formation of each species.

The differential rate equations applicable to the previously deduced simplified reaction scheme, (B), are

$$\frac{dC_{NA}}{dt} = -(k_1 + k_2)\, C_{NA} = -2\, k_1 C_{NA} \tag{7.2}$$

$$\frac{dC_{NQ}}{dt} = k_1\, C_{NA} - k_3\, C_{NQ} \tag{7.3}$$

$$\frac{dC_{PA}}{dt} = k_2\, C_{NA} + k_3\, C_{NQ} - k_4\, C_{PA} \tag{7.4}$$

where C_{NA}, C_{PA}, C_{NQ} are the concentrations of naphthalene, phthalic anhydride and naphthoquinone after any time, t, spent in the reactor.

Integration of the first order equation (7.2) leads to

$$C_{NA} = C_{NA_0}\, e^{-2k_1 t}. \tag{7.5}$$

Hence substitution in equation (7.3) results in the following first order linear differential equation

$$\frac{dC_{NQ}}{dt} + k_3\, C_{NQ} = k_1\, C_{NA_0}\, e^{-2k_1 t}. \tag{7.6}$$

As has been seen previously in Chapter 2, the solution to this equation is of the form

$$y\, e^{\int P dx} = \int Q\, e^{\int P dx}\, dx + \text{constant}.$$

The integrating factor to be applied in the solution of equation (7.6) is $e^{k_3 t}$. The solution to this equation is then deduced to be

$$\frac{C_{NQ}}{C_{NA_0}} = \frac{k_1}{k_3 - 2k_1} \{e^{-2k_1 t} - e^{-k_3 t}\}. \tag{7.7}$$

Similarly, substitution for C_{NA} and C_{NQ} in equation (7.4) gives

$$\frac{dC_{PA}}{dt} + k_4 C_{PA} = k_1 C_{NA_0} e^{-2k_1 t} + \frac{k_3 k_1 C_{NA_0}}{k_3 - 2k_1} \{e^{-2k_1 t} - e^{-k_3 t}\}.$$

$$(7.8)$$

This equation is integrated using the factor $e^{k_4 t}$ and the constant term is evaluated from the condition that at $t = 0$, $C_{PA} = 0$.

The solution to equation (7.8) is

$$\frac{C_{PA}}{C_{NA_0}} = \frac{k_1}{(k_4 - 2k_1)} \{e^{-2k_1 t} - e^{-k_4 t}\} + k_3 k_1$$

$$\left\{ \frac{e^{-2k_1 t}}{(k_3 - 2k_1)(k_4 - 2k_1)} + \frac{e^{-k_3 t}}{(2k_1 - k_3)(k_4 - k_3)} \right.$$

$$\left. + \frac{e^{-k_4 t}}{(k_4 - k_3)(k_4 - 2k_1)} \right\}.$$

$$(7.9)$$

For a reactor considered to be a continuous stirred tank, i.e. with complete mixing, the product concentrations are determined by carrying out mass balances across the reactor for each species in the usual fashion.

Applying a mass balance to naphthalene we obtain

$$v C_{NA_0} = v C_{NA} + (-r_{NA}) V = v C_{NA} + (2k_1 C_{NA}) V. \qquad (7.10)$$

Dividing throughout by $v C_{NA_0}$ we get

$$1 = \frac{C_{NA}}{C_{NA_0}} + \frac{2k_1 C_{NA} \bar{t}}{C_{NA_0}}$$

where $\bar{t}$ is mean residence time in reactor,

or $$\frac{C_{NA}}{C_{NA_0}} = \frac{1}{(1 + 2k_1 \bar{t})}. \qquad (7.11)$$

Similarly, by applying a mass balance to naphthoquinone we get

$$0 = v C_{NQ} + (-r_{NQ}) V = v C_{NQ} + (k_3 C_{NQ} - k_1 C_{NA}) V \qquad (7.12)$$

where $C_{NQ_0} = 0$. Substitution for C_{NA} and rearrangement gives

$$\frac{C_{NQ}}{C_{NA_0}} = \frac{k_1 (C_{NA}/C_{NA_0}) \bar{t}}{(1 + k_3 \bar{t})} = \frac{k_1 \bar{t}}{(1 + k_3 \bar{t})(1 + 2k_1 \bar{t})}. \qquad (7.13)$$

Finally, from a mass balance on phthalic anhydride, the concentration of phthalic anhydride may be calculated

$$0 = vC_{PA} + (-r_{PA})V = vC_{PA} + (k_4C_{PA} - k_1C_{NA} - k_3C_{NQ})V. \quad (7.14)$$

Substitution for C_{NA}, C_{NQ} and rearrangement results in

$$\frac{C_{PA}}{C_{NA_0}} = \frac{\{k_1(C_{NA}/C_{NA_0}) + k_3(C_{NQ}/C_{NA_0})\}\bar{t}}{(1+k_4\bar{t})}$$

$$= \left\{\frac{k_1\bar{t}}{(1+2k_1\bar{t})} + \frac{k_3k_1\bar{t}^2}{(1+k_3\bar{t})(1+2k_1\bar{t})}\right\} \frac{1}{(1+k_4\bar{t})}$$

$$= \frac{k_1\bar{t}}{(1+k_4\bar{t})(1+2k_1\bar{t})} \left\{1 + \frac{k_3\bar{t}}{(1+k_3\bar{t})}\right\}. \quad (7.15)$$

De Maria *et al.* (1961), by solving equations which incorporated a degree of mixing term, obtained concentration data for phthalic anhydride and naphthoquinone as a function of temperature and contact time. Data obtained using their catalyst A are reproduced in Fig. 7.1. The figures on this graph refer to the value of the Peclet number, which is the ratio of the effective diffusion time, $h^2/2D_e$, to the average residence time, h/u ($Pe = uh/2D_e$ where u is the true axial velocity in the bed, h is the bed height and D_e is the effective diffusivity). For plug flow $Pe = \infty$ while for complete mixing $Pe = 0$. The above equations, (7.7), (7.9), (7.13) and (7.15), may therefore be used to calculate the expected naphthoquinone and phthalic anhydride concentrations at these extremes of behaviour. An industrial fluidised bed reactor has a Peclet number between 0·2 and 3, i.e. closer to the stirred tank model than to the plug flow.

It is seen from Fig. 7.1 that a temperature of about 623°K will result in maximum anhydride concentration for a contact time of six seconds, but operation should take place at a still higher temperature to reduce the amount of naphthoquinone produced.

Fig. 7.2 shows how the conversions to phthalic anhydride and naphthoquinone vary with temperature and Peclet number at contact times of 6 and 18 seconds. It will be seen that for operation at the longer contact time of 18 seconds (near the published industrial operating residence time), the optimum temperature (603°K) is lower than is the case at 6 seconds (623°K), but a higher temperature is desirable for suppression of naphthoquinone.

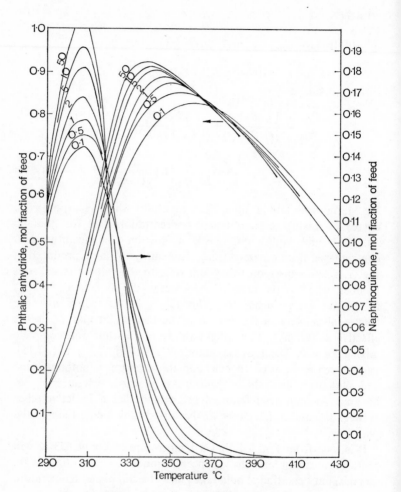

FIG. 7.1. Yield of phthalic anhydride and naphthoquinone as a function of temperature. (Numbers on curves refer to values of Peclet number) (Reprinted from *Ind. Eng. Chem.*, **53**, April 1961, p. 259. Copyright (1961) by the American Chemical Society, Reprinted by permission of the copyright owner)

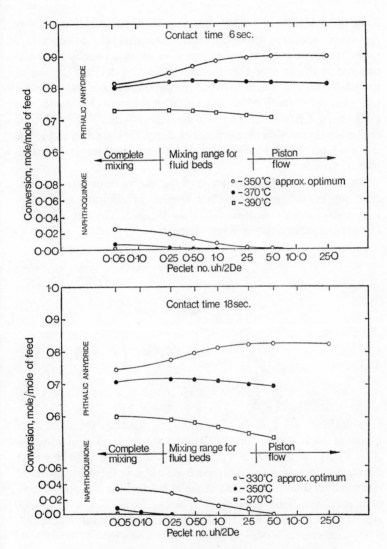

FIG. 7.2. Conversion as a function of Peclet number and
contact time (Permission granted as in Fig. 7.1.)

7.2.5 ANALYSIS OF NAPHTHALENE OXIDATION IN A FIXED BED REACTOR

Since the oxidation of naphthalene to phthalic anhydride is a highly exothermic reaction, it might be anticipated that severe temperature gradients could occur radially as well as axially, thereby necessitating the application of the two-dimensional model and the use of partial differential equations to predict the temperature and composition changes in both axial and radial directions. This model was outlined in Chapter 2. The one-dimensional model will also be applied in this section for purposes of comparison, and to emphasise the need to consider concentration and temperature gradients in both directions.

We are dealing here with a complex set of four simultaneous reactions involving four reacting species, in addition to oxygen. Four mass balance equations are thus required. The rate equations for reaction scheme (B) have been given as

$$\frac{dC_{NA}}{dt} = -(k_1 + k_2)C_{NA} \tag{7.2}$$

$$\frac{dC_{NQ}}{dt} = k_1 C_{NA} - k_3 C_{NQ} \tag{7.3}$$

$$\frac{dC_{PA}}{dt} = k_2 C_{NA} + k_3 C_{NQ} - k_4 C_{PA} \tag{7.4}$$

$$\frac{dC_{MA}}{dt} = k_4 C_{PA}. \tag{7.16}$$

Now, based on 1 mole of naphthalene reacting, the number of moles of each species present will be $n_{NA} = 1 - X_1 - X_2$, $n_{NQ} = X_1 - X_3$, $n_{PA} = X_2 + X_3 - X_4$, where X_1, X_2, X_3 and X_4 represent conversion in reactions 1–4 respectively. If ρ is the gas density, as measured by total moles/free volume of reactor, the concentration of each species will be given by $C_{NA} = (n_{NA}/n_T)\rho$, $C_{NQ} = (n_{NQ}/n_T)\rho$, etc., n_T being the total number of moles per mole of naphthalene fed.

The concentration of naphthalene is then given by

$$C_{NA} = \frac{(1 - X_1 - X_2)\rho}{n_T}, \tag{7.17}$$

and if ρ/n_T is a weak function of time, as would be the case since the air/naphthalene ratio is very high,

$$\frac{dC_{NA}}{dt} = \frac{\rho}{n_T}\left(-\frac{dX_1}{dt} - \frac{dX_2}{dt}\right). \tag{7.18}$$

But from equation (7.2)

$$\frac{dC_{NA}}{dt} = -(k_1 + k_2)\,C_{NA},$$

and this may be written as

$$\frac{dC_{NA}}{dt} = -(R_1 + R_2) \tag{7.19}$$

where R_1 is the rate of reaction in step 1 and R_2 is rate in step 2 of scheme (B). The units of R_1 and R_2 are mol/(free volume of reactor) (time). Hence if these reactions are considered to occur independently of each other, but in parallel

$$R_1 = \frac{\rho}{n_T}\frac{dX_1}{dt} \tag{7.20}$$

$$R_2 = \frac{\rho}{n_T}\frac{dX_2}{dt}. \tag{7.21}$$

Now, for a tubular reactor with plug flow,

$$dt = \frac{dV}{v} = \frac{dZ\,A_c\,\varepsilon}{(F_{NA}n_T)/\rho} \tag{7.22}$$

where v is volumetric flow rate of gas, A_c is the cross-sectional area of reactor and ε is the voidage of the catalyst bed.

Hence, by substituting equation (7.22) in equations (7.20) and (7.21), we obtain

$$R_1 = \frac{\rho}{n_T}\frac{dX_1\,F_{NA}\,n_T}{dZA_c\,\varepsilon\rho} \quad \text{and} \quad R_2 = \frac{\rho}{n_T}\frac{dX_2\,F_{NA}\,n_T}{dZA_c\,\varepsilon\rho},$$

or

$$\frac{F_{NA}dX_1}{dZ} = R_1\,A_c\,\varepsilon \tag{7.23}$$

and

$$\frac{F_{NA}\,dX_2}{dZ} = R_2\,A_\bullet\,\varepsilon. \tag{7.24}$$

Similarly

$$\frac{F_{NA}\,dX_3}{dZ} = R_3\,A_c\,\varepsilon \quad (7.25) \quad \text{and} \quad \frac{F_{NA}\,dX_4}{dZ} = R_4\,A_c\,\varepsilon \quad (7.26)$$

where $R_3 = k_3\,C_{NQ}$ and $R_4 = k_4 C_{PA}$.

The mass balance equations are now presented as equations involving conversions in a particular step.

The energy balance equation can also be written in terms of the sum of the heats evolved in each of the reaction steps

$$F_{NA}\sum_{i=1}^{4}(-\Delta H_i)\frac{dX_i}{dZ} - UA_w(T_m - T) = F_{NA}\sum_{j=1}^{w} n_j C_{pj}\frac{dT}{dZ} \quad (7.27)$$

where X_i is conversion in reaction i, having heat of reaction $(-\Delta H_i)$, and n_j is moles of species j/mole of naphthalene of heat capacity C_{pj}.

This one-dimensional model assumes that the temperature and composition are uniform across the diameter of the tube, and that the temperature shows a sharp drop at the wall where all the resistance to heat transfer is assumed to occur.

Using the rate constant data of de Maria et al. (1961), calculated heats of reaction and specific heat data, and a value of the wall heat transfer coefficient calculated according to the correlation of Leva (1948), the above equations were solved (Parsons, 1970).

The general method of solution of this type of problem was outlined in Chapter 2. A machine solution to this system of differential equations was obtained by the use of a Continuous System Modelling Programme (C.S.M.P). The output values are conversions in reactions 1–4 and the temperature at points along the reactor.

The two-dimensional model can be applied at different degrees of complexity. Because the reactor tube contains both gas and solid catalyst, the most rigorous approach would involve postulating equations for the heat effects in the two phases separately. At the present stage of development such a model cannot be applied and, in any case, bearing in mind the accuracy of the experimental data, its application is probably not justified. Instead the system is assumed to consist of a homogeneous mass within which gas diffusion is governed by an effective diffusivity, D_e, and heat is transferred by a form of conduction described by an effective thermal conductivity, k_e, as discussed in Chapter 2. Diffusion in an axial direction can be neglected in relation to transport by bulk flow and a constant gas velocity is assumed across the tube diameter, the gas flow then obeying a plug flow model. The effective thermal conductivity may be assumed constant across the tube diameter although there is evidence that this has a much smaller value at the tube wall.

Forms of the differential equations required were developed in Chapter 2 (equations 2.60 and 2.62).

Mass balance

The mass balance equations are written in terms of conversions occurring in each separate reaction, as in the one-dimensional case. The form of the equation is

$$\frac{\partial X_i}{\partial Z} - \frac{D_e}{\varepsilon u}\left(\frac{\partial^2 X_i}{\partial r^2} + \frac{1}{r}\frac{\partial X_i}{\partial r}\right) - \frac{R_i \varepsilon}{G} = 0 \tag{7.28}$$

where R_i represents the rate of reaction in step i and u is the velocity based on cross-sectional area of the tube. There are four steps in the reaction scheme so four mass balance equations are required.

Energy balance

$$\frac{\partial T}{\partial Z} - \frac{k_e}{G \sum\limits_{j=1}^{w}(X_j C_{p_j})}\left(\frac{\partial^2 T}{\partial r^2} + \frac{1}{r}\frac{\partial T}{\partial r}\right) + \frac{\varepsilon \sum\limits_{i=1}^{4}(-\Delta H_i)R_i}{G \sum\limits_{j=1}^{w}(X_j C_{p_j})} = 0 \tag{7.29}$$

where GC_p is replaced by $G \sum\limits_{j=1}^{w}(X_j C_{p_j})$, G being flow of naphthalene in feed/cross-sectional area and X_j is conversion of species j of specific heat C_{p_j}. The rate has been transformed into that for reaction i.

The heat and mass balance equations are solved in the general fashion outlined in Chapter 2. The partial differential equations are transformed into difference equations which are solved by machine computation, forward difference formulae being used to initialise the problem from the inlet position, and central differences in the remaining part of the simulation.

Thus, for the temperature distribution initially,

$$T_{m,n+1} = T_{m,n} + \frac{k_e}{P}\frac{\Delta Z}{(\Delta r)^2}\left[\frac{1}{2m}(T_{m+1,n} - T_{m-1,n}) + T_{m+1,n}\right.$$
$$\left. - 2T_{m,n} + T_{m-1,n}\right] + \frac{S\Delta Z}{P} \tag{7.30}$$

where $P = G \sum_{j=1}^{w} X_j C_{p_j}$ and $S = \varepsilon \sum_{i=1}^{4} (-\Delta H_i) R_i$,

and for the remainder,

$$T_{m,n+1} = T_{m,n-1} + \frac{2k_e}{P} \frac{\Delta Z}{(\Delta r)^2} \left[\frac{1}{2m} (T_{m+1,n-1} - T_{m-1,n-1}) \right.$$

$$\left. + T_{m+1,n-1} - 2T_{m,n-1} + T_{m-1,n-1} \right] + \frac{S\Delta Z2}{P}. \qquad (7.31)$$

In a similar fashion equations for conversion can be written.

The average conversion across the tube diameter is obtained by application of equation (2.75).

The results obtained by application of the two models (Parsons, 1970) will now be considered. As far as possible functions were kept the same in each. For example the same values of catalyst pellet diameter, voidage, linear velocity of gas, rate data, heats of reaction and heat capacities were used in the two cases. It is necessary of course in the complex model to include values for D_e and k_e which do not occur in the simple model. In the latter an overall heat transfer coefficient, U, has been introduced but this is not comparable with k_e; Froment (1967) has shown how the overall heat transfer coefficient may be related to the parameters of the two-dimensional model. The pressure was assumed constant at 1 atm. and the inlet temperature constant across the tube, except in the complex model where the temperature at all radial positions had the inlet value but the wall temperature was maintained constant at 620°K.

The effects of the following variables on the conversion and temperature profiles were examined: air/naphthalene ratio, inlet temperature, tube diameter and, in the case of the two-dimensional model, k_e and D_e were also varied.

The air/naphthalene ratio was shown to be important in controlling the temperature of operation. As can be seen from Fig. 7.3 for the one-dimensional case, if the ratio becomes too small, excessive temperatures are generated along the reactor length, the temperature being limited only by the fact that all naphthoquinone is used up, halting step 3 which is faster than the others. Similarly, as shown in Fig. 7.4, increasing the tube diameter to 50 mm instead of 30 mm also results in excessive temperatures being attained in the tube for the particular chosen values of the other parameters. The extreme sensitivity of reactor temperature to variation in entrance temperature

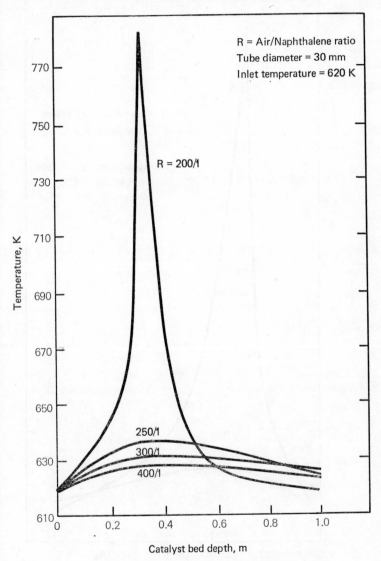

FIG. 7.3. Temperature variation with air/naphthalene ratio
—simple model

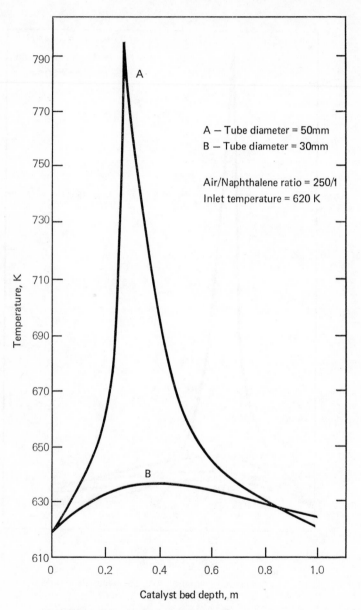

A — Tube diameter = 50mm
B — Tube diameter = 30mm

Air/Naphthalene ratio = 250/1
Inlet temperature = 620 K

FIG. 7.4. Temperature profiles for tube diameters of 30 and
50 mm—simple model

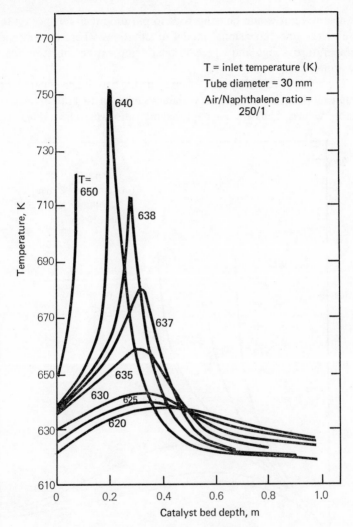

FIG. 7.5. Temperature variation along catalyst bed as a function of gas inlet temperature—simple model

is shown in Fig. 7.5. This phenomenon, known as parametric sensitivity, is a well documented effect. These results for the one-dimensional case apply for an average radial temperature, and the temperature produced at the tube centre-line will be greater than those shown. Thus, in view of the extreme sensitivity to entrance temperature, it would be dangerous to put too much reliance on the use of the one-dimensional model in situations where the entrance temperature is high and where 'critical' temperature conditions may be approached.

The effect of variation of k_e when using the two-dimensional model was examined by setting the variables common to both models at stable values. Thus the air/naphthalene ratio was set at 250/1, the

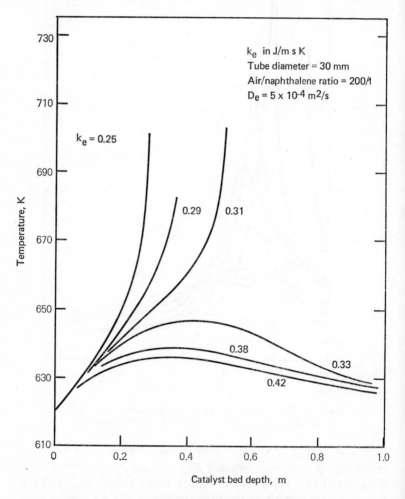

FIG. 7.6. Core temperature variation with thermal conductivity of catalyst bed—complex model

tube diameter at 30 mm and the entrance temperature at 620°K. The effective diffusivity was set first at $5 \times 10^{-4} \, m^2/s$. It can be seen from Fig. 7.6 that the effect of decrease in k_e on stability of operation is very marked. Low thermal conductivities help to build up large radial temperature gradients. Variation of D_e from 5 to 10×10^{-4}

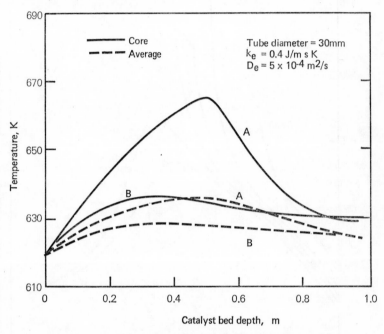

FIG. 7.7. Core and average temperatures—complex model
A—air/naphthalene ratio = 200/1
B—air/naphthalene ratio = 250/1

m^2/s had very little effect on the temperature profile suggesting that the one-dimensional model gives an adequate description of the mass continuity equations. The centre-line and average temperatures are shown as a function of bed depth in Fig. 7.7 for two different reactant ratios. For the lower ratio case the difference in values at equivalent bed depths is seen to be large, indicating substantial radial gradients. The extreme sensitivity to entrance temperature of bed temperature for this reactant ratio is seen in Fig. 7.8. A change from 621–622°K is enough to cause temperature runaway. In general the one-dimen-

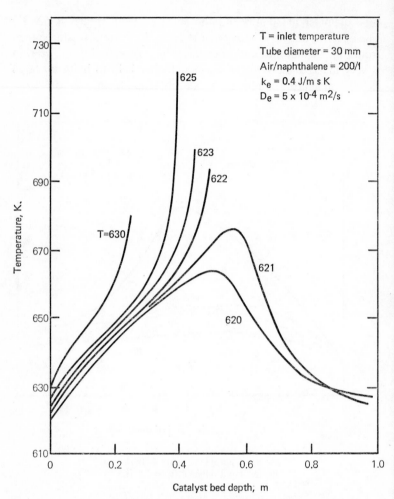

FIG. 7.8. Influence of inlet temperature on stability of reaction, showing the core temperature for the complex model

sional model predicts higher conversions than does the two. This is because the two-dimensional model assumes a constant wall temperature which is of course lower than the temperature within the tube, thus lowering the average temperature in the tube even though the core temperature is higher. This is illustrated in Fig. 7.9 where, for a starting temperature of $640°K$, the one-dimensional case points to instability while the two is quite stable. However it must be remembered that considerable uncertainty exists because the heat transfer parameters are not exactly equivalent.

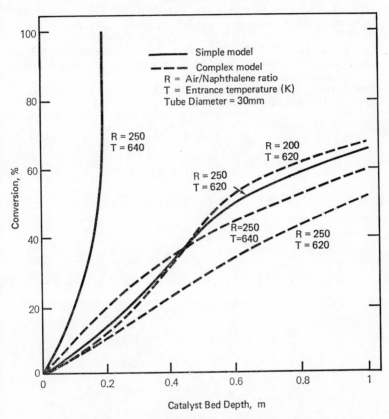

FIG. 7.9. Conversion and average conversion as a function of catalyst bed depth for the simple and complex models respectively

For the two-dimensional case, with an air/naphthalene ratio of 200/1 and a 30 mm diameter tube, an inlet temperature of 620°K, $D_e = 5 \times 10^{-4}\,\text{m}^2/\text{s}$ and $k_e = 0.4\,\text{J/m s °K}$, a conversion of 83 % to phthalic anhydride is calculated for a tube length of 2 m.

In this discussion no account is taken of inter and intra phase mass transfer effects.

REFERENCES

CARBERRY, J. J. and WHITE, D. 1969. *Ind. Eng. Chem.* **61**(7), 27.

FROMENT, G. F. 1967. *Ind. Eng. Chem.* **59**(2), 18.

GREEN, J. H. S. 1961. *Quart. Rev.* **15**, 125.

LEVA, M., GRUMMER, M. and CLARK, E. L. 1948. *Ind. Eng. Chem.* **40**, 747.

DE MARIA, F., LONGFIELD, J. E. and BUTLER, G. 1961. *Ind. Eng. Chem.* **53**, 259.

PARSONS, D. D. 1970. M.Sc. dissertation, Univ. of Newcastle upon Tyne.

REID, R. C. and SHERWOOD, T. K. 1966. *The properties of gases and liquids*, 2nd Ed. McGraw-Hill, New York.

RUTHRUFF, R. F. 1953. *Pet. Ref.* **32**(10), 113.

SPITZ, P. H. 1968. *Hydro. Proc.* **47**(11), 162.

8

HOMOGENEOUS CATALYSIS BY CO-ORDINATION COMPOUNDS OF TRANSITION METALS—WACKER PROCESS FOR ACETALDEHYDE SYNTHESIS

8.1 Introduction

In earlier chapters the application of heterogeneous catalysis to chemical synthesis on an industrial scale has been amply demonstrated. Processes involving gaseous reaction on solid catalysts have been operated successfully for more than fifty years despite the fact that the function of these catalysts is still only partially understood. In contrast chemical reactions, catalysed by substances dissolved in a liquid phase, are a more recent innovation but already many have been developed commercially, and processes utilising homogeneous catalysts are now well established. The rapid expansion in the exploitation of this type of process has come about because of the great versatility and ease of application of the co-ordination compounds of the transition metals as catalysts in solution. These processes are on the whole more reproducible and simpler in nature than heterogeneous processes, and a good understanding of the mechanism of many of these reactions has already been obtained. Table 8.1 gives details of important industrial reactions which are homogeneously catalysed.

8.2 Homogeneous Catalysis by Co-ordination Compounds of Transition Metals

The homogeneous catalysts employed are co-ordination compounds mainly of the group VIII elements—Fe, Co, Ni, Ru, Rh, Pd, Os, Ir,

TABLE 8.1

Process Summary (From Prengle and Barona, 1970a)

(1) Process/Reaction	(2) Raw Materials	(3) Approximate Conditions	(4) Catalysts	(5) Yield and Selectivity	(6) Rate Controlling Step	(7) Reactor Configuration
(1) Acetaldehyde, CH_3CHO (Hoechst–Wacker Process) $C_2H_4 + 2CuCl_2 + H_2O \xrightarrow{PdCl_2}$ $CH_3CHO + 2HCl + 2CuCl$ $2CuCl + 2HCl + \frac{1}{2}O_2 \rightarrow 2CuCl_2 + H_2O$	C_2H_4 O_2 (Air) Catalysts	100°C (373°K) 6 atm. Liquid-Gas Phase H_2O Solvent pH 0.8–3.0 Also neutral media	Homogeneous liquid phase $PdCl_2$ $CuCl_2$	>95%, ≈96%	Mass Transfer of C_2H_4 & O_2	1 or 2 stages sparged, CSTR or unstirred
(2a) Acetic Acid, CH_3COOH (Celanese; Chemische Werke Huels) $n-C_4H_{10} + 5/2O_2 \rightarrow$ $2CH_3COOH + H_2O$ (Distillers Co., C_4-C_7)	C_4 or C_4-C_7 O_2 (Air) Catalysts	150–225°C (423–498°K) 55 atm. Liquid-Gas Phase Solvents: C_1-C_{10} aliphatic acids	Homogeneous Cr, Mn, Co, Ni-Acetates, Propionates, Butyrates, Naphthenates	≈93%, ≈45%	Mass Transfer of O_2	Sparged
(2b) Acetic Acid, CH_3COOH (GfGE-Shawinigan) $CH_3CHO + \frac{1}{2}O_2 \rightarrow CH_3COOH$	Acetaldehyde O_2 (Air)	55–65°C (328–338°K) 6 atm. Acetic Acid as Solvent	Homogeneous Manganous Acetate, Cobalt Acetate	≈95%	Mass Transfer of O_2	Oxidation tower, sparged at multiple levels, w/side stream heat removal
(3) Acetone, $(CH_3)_2CO$ (Wacker-Hoechst Process) $C_3H_6 + 2CuCl_2 + H_2O \xrightarrow{PdCl_2}$ $(CH_3)_2CO + 2HCl + 2CuCl$ $2CuCl + 2HCl + \frac{1}{2}O_2 \rightarrow 2CuCl_2 + H_2O$	C_3H_6 O_2 (Air) Catalysts	50–120°C (323–393°K) 50–100 atm. Liquid-Gas Phase H_2O Solvent pH 0.7–3.0 or 2–2.6	Homogeneous Liquid Phase $Pd\,Cl_2$ $CuCl_2$	≈99%	Mass Transfer of C_3H_6 and O_2	1 or 2 stages sparged, CSTR or vertical tubes, continuous
(4a) Adipic Acid, $(C_2H_4COOH)_2$ (Du Pont Process) $C_6H_{11}OH + 2HNO_3 \rightarrow$ $C_6H_{10}O_4 + 2H_2O + N_2O$	Cyclohexanol & Cyclohexanone HNO_3 Catalyst	75–85°C (348–358°K) 3 atm.	Homogeneous 0-2 wt.% NH_4VO_3 + $Cu(NO_3)_2$	≈92–94%	Mass Transfer of O_2	CSTR sparged

TABLE 8.1 (continued)

(1) Process/Reaction	(2) Raw Materials	(3) Approximate Conditions	(4) Catalysts	(5) Yield and Selectivity	(6) Rate Controlling Step	(7) Reactor Configuration
(4b) Adipic Acid, $(C_2H_4COOH)_2$ (I. G. Farben Process) $C_6H_{11}OH + 2O_2 \rightarrow C_6H_{10}O_4 + H_2O$	Cyclohexanol & Cyclohexanone O_2 (Air) Catalyst	75–85°C (348–358°K) 1 atm.	Homogeneous 0·05–0·07 wt. % Mn Acetate + 0·125–0·175 wt. % Ba Acetate	90%	Mass Transfer of O_2	CSTR sparged
(5) Benzoic Acid, C_6H_5COOH (Mid-Century Process) $C_6H_5CH_3 + 3/2O_2 \rightarrow C_6H_5COOH + H_2O$	Toluene O_2 (Air) Catalyst	200°C (473°K) 28 atm.	Homogeneous 0·9 wt. % tetrabromomethane + 0·9 wt. % Co & Mn Acetates	≃90–95%	Mass Transfer of O_2	Agitated sparged, heat removed by vaporization or circulation
(6) Butyric Acid, C_3H_7COOH (Commercial Solvents Process) $3C_3H_7CHO + 3/2O_2 \rightarrow C_3H_7COOH + (C_3H_7CO)_2 + H_2O$	Butyraldehyde O_2 (Air) Catalyst	65–75°C (338–348°K) 6–7 atm.	Homogeneous 0·5 wt. % Mn butyrate	≃95%	Mass Transfer of O_2	Sparged at multiple levels with side stream heat removal
(7) Cyclohexanol & Cyclohexanone, $C_6H_{11}OH$ $C_6H_{10}O$ (Du Pont) $C_6H_{12} + ((1+x)/2)O_2 \rightarrow x(C_6H_{10}O) + (1-x)C_6H_{11}OH + xH_2O$	C_6H_{12} O_2 (Air) Catalyst	150–160°C (423–433°K) 8–10 atm.	0·005 wt. % Co(III) naphthenate Scavenger-H_3BO_3 or B_2O_3	≃60–90% ≃70–79%	Mass Transfer of O_2	Multiple continuous sparged or staged reactor
(8) Methyl Isobutyl Ketone, (British Petroleum Process) $(CH_3)_2C = C_3H_6 + \frac{1}{2}O_2 \rightarrow$ CHCOCH$_3$ $(CH_3)_2C = CHCOCH_3 + H_2O$ $(CH_3)_2C = CHCOCH_3 + H_2 \rightarrow$ $C_4H_9COCH_3$	Mixed Methyl Pentenes O_2 (Air) Catalyst	45–75°C (318–348°K) 10 atm. Solvent: reactant mixture	Cobalt Naphthenate (0·2 g/l)	≃90%	Mass Transfer of O_2	Sparged CSTR or unstirred

TABLE 8.1 (*continued*)

(1) Process/Reaction	(2) Raw Materials	(3) Approximate Conditions	(4) Catalysts	(5) Yield and Selectivity	(6) Rate Controlling Step	(7) Reactor Configuration
(9) Phenol, C_6H_5OH (Dow Process) $C_6H_5COOH + \frac{1}{2}O_2 \rightarrow C_5OH + CO_2$	C_6H_5COOH O_2 (Air) Steam Catalyst	230–240°C (503–513°K) 1 atm.	1–10 wt. % Copper benzoate promoted with Mg Benzoate	≃89–93%	Mass Transfer of O_2	CSTR sparged
(10a) Terephthalic Acid, 1-C_6H_4(COOH)$_2$ (Mid-Century Process) p-$C_6H_4(CH_3)_2 + 3O_2 \rightarrow$ p-$C_6H_4(COOH)_2 + 2H_2O$	p-Xylene O_2 (Air) Catalyst	195–205°C (468–478°K) 26 atm. Solvent: CH_3COOH	0–45 wt. % Co, or Mn acetate or NH_4 molybdate promoted with bromine	90%	Mass Transfer of O_2	Sparged
(10b) Terephthalic Acid, p-C_6H_4(COOH)$_2$ (Teijin Process) p-$C_6H_4(CH_3)_2 + 3O_2 \rightarrow$ p-$C_6H_4(COOH)_2 + 2H_2O$	p-Xylene O_2 (Air) Catalyst	100–130°C (373–403°K) 10 atm. Solvent: CH_3COOH	20–100 wt. % Co-based catalyst	97–98%	Mass Transfer of O_2	CSTR sparged
(11) Vinyl Acetate, $C_2H_3COOCH_3$ (Hoechst Process) $C_2H_4 + 2CuCl_2 \xrightarrow{\;CH_3COOH\;\; PdCl_2\;}$ $C_2H_3COOCH_3 + 2HCl + 2CuCl$ $2CuCl + 2HCl + \frac{1}{2}O_2 \rightarrow 2CuCl_2$ $+ H_2O$	C_2H_4 O_2 (Air) Catalyst	100–130°C (373–403°K) 30–40 atm.	Homogeneous $PdCl_2$ $CuCl_2$	≃90%	Mass Transfer of C_2H_4 & O_2	1 or 2 stages, sparged, CSTR or unstirred

Pt—although certain transition metals in groups IVB–VIIB and Cu, Ag and Au in group IB also form complexes which can act as catalysts. These metals owe their unusual properties to unfilled inner electron shells, or, more specifically, the d-orbitals in which the electron configurations are generally in the range d^6 to d^{10}. The electron configurations of the Group VIB–VIII and IB elements are shown in Table 8.2. These metals of course are capable of forming positive ions by loss of electrons as shown in Table 8.3, but in addition have the capability of accepting lone pairs of electrons from donor groups or ligands into the d, s and p orbitals which have rather similar energy levels. These orbitals in fact undergo hybridisation

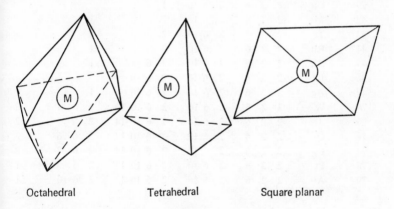

Octahedral   Tetrahedral   Square planar

FIG. 8.1. Shape of metal complexes

which results in an equalisation of energy associated with the orbitals and a strengthening as compared with the simple orbitals alone. By appropriate mixing d^2sp^3, dsp^2 and sp^3 hybridisation is possible, the type of hybridisation being associated with a characteristic geometric configuration. Thus the d^2sp^3 combination gives rise to a regular octahedral configuration, the dsp^2 a square planar arrangement and the sp^3 a tetrahedral (Fig. 8.1). The co-ordination number in these cases will be 6, 4 and 4 respectively indicating the number of orbitals being filled by pairs of donor electrons. The formation of ammine complexes of cobalt, nickel and zinc, which are octahedral, square planar and tetrahedral respectively

TABLE 8.2

Electron configuration of atoms of groups VIB, VIIB, VIII and IB

Atomic number	Element	Main shell 1 s	2 s	2 p	3 s	3 p	3 d	4 s	4 p	4 d	4 f	5 s	5 p	5 d	5 f	6 s
24	Cr	2	2	6	2	6	5	1								
25	Mn	2	2	6	2	6	5	2								
26	Fe	2	2	6	2	6	6	2								
27	Co	2	2	6	2	6	7	2								
28	Ni	2	2	6	2	6	8	2								
29	Cu	2	2	6	2	6	10	1								
42	Mo	2	2	6	2	6	10	2	6	5		1				
43	Tc	2	2	6	2	6	10	2	6	5		2				
44	Ru	2	2	6	2	6	10	2	6	7		1				
45	Rh	2	2	6	2	6	10	2	6	8		1				
46	Pd	2	2	6	2	6	10	2	6	10						
47	Ag	2	2	6	2	6	10	2	6	10		1				
74	W	2	2	6	2	6	10	2	6	10	14	2	6	4		2
75	Re	2	2	6	2	6	10	2	6	10	14	2	6	5		2
76	Os	2	2	6	2	6	10	2	6	10	14	2	6	6		2
77	Ir	2	2	6	2	6	10	2	6	10	14	2	6	7		2
78	Pt	2	2	6	2	6	10	2	6	10	14	2	6	9		1
79	Au	2	2	6	2	6	10	2	6	10	14	2	6	10		1

TABLE 8.3

Electronic distribution of some transition metal ions

		$3d$	$4s$	$4p$
3+ ions	Cr	(↑)(↑)(↑)( )( )	( )	( )( )( )
	Mn	(↑)(↑)(↑)(↑)( )	( )	( )( )( )
	Fe	(↑)(↑)(↑)(↑)(↑)	( )	( )( )( )
	Co	(↑↓)(↑)(↑)(↑)(↑)	( )	( )( )( )
2+ ions	Ni	(↑↓)(↑↓)(↑↓)(↑)(↑)	( )	( )( )( )
	Cu	(↑↓)(↑↓)(↑↓)(↑↓)(↑)	( )	( )( )( )
	Zn	(↑↓)(↑↓)(↑↓)(↑↓)(↑↓)	( )	( )( )( )

are shown as follows:

$Co(NH_3)_6^{3+}$

3d	4s	4p
$|1\downarrow|1\downarrow|1\downarrow|xx|xx|$	$|xx|$	$|xx|xx|xx|$

d^2sp^3 hybridisation

$Ni(NH_3)_4^{2+}$

3d	4s	4p
$|1\downarrow|1\downarrow|1\downarrow|1\downarrow|xx|$	$|xx|$	$|xx|xx|\ \ |$

dsp^2 hybridisation

$Zn(NH_3)_4^{2+}$

3d	4s	4p
$|1\downarrow|1\downarrow|1\downarrow|1\downarrow|1\downarrow|$	$|xx|$	$|xx|xx|xx|$

sp^3 hybridisation

(1 indicates electron associated with metal nucleus and x a donated electron).

It may be seen from the structures above that pairing of the $3d$ electrons has taken place in the case of the cobalt and nickel metals but this does not universally occur. Ligand field theory can predict whether pairing will take place or not and, putting it very simply, the stronger the tendency of the ligand to donate electrons (for example, NH_3 or CN^-), the stronger the field due to the approaching groups, and the greater the tendency for pairing of the central ion's electrons to take place. This results in a low spin or spin paired complex. A weakly donating ligand will give rise to a high spin complex, the electrons of the central atom tending to maintain parallel spins as far as possible, as shown in Table 8.3. Configurations, such as $Co(NH_3)_6^{3+}$, where the valence shell of the metal atom contains 18 electrons are particularly stable, those of $Mn(NH_3)_6^{3+}$ and $Fe(NH_3)_6^{3+}$ which have 16 and 17 respectively slightly less so.

The importance of transition metal complexes in homogeneous catalysis can be related to the following factors (Halpern, 1968).

(a) Among the ligands which can be co-ordinated are many organic compounds including alkenes, alkynes, aromatic hydrocarbons, aromatic nitrogen molecules and many organic molecules containing

H

oxygen, including alcohols, phenols, ethers and acids. This fact has obvious significance for the synthesis of organic materials.

(b) A catalyst may function by providing a route to product which requires the expenditure of less energy than in the non-catalytic route. The formation of an intermediate complex as a step in this route can constitute a low energy path. For example, hydrogenation reactions can be effected by reactive atomic ions brought about by the heterolytic splitting of H_2 according to $H_2 \rightarrow H^+ + H^-$ which requires an estimated activation energy of 146 kJ/mol but a catalytic route requiring less energy involves the reaction

$$Ru^{III}Cl_6^{3-} + H_2 \rightarrow Ru^{III}Cl_5H^{3-} + H^+ + Cl^-. \qquad (8.1)$$

The transition metal here stabilises, by σ bonding, an otherwise reactive intermediate, H^-. Alkyl groups can be stabilised in the same way, as also can the alkenes, as already mentioned, but this time by π bonding.

(c) Many of the complexes formed, which are stable, are also highly reactive, being analogous to the reactive intermediates of organic chemistry viz. free radicals ($R_3C\cdot$), carbenes ($R_2C:$), carbonium ions (R_3C^+) and carbanions ($R_3C:^-$). The analogy between the species and corresponding transition metal counterparts is depicted in Table 8.4.

The reactive intermediate complex analogous to a free radical, for example $Co(CN)_5^{3-}$, is pictured as being formed from the very stable d^6 complex as follows:

$$Co(CN)_6^{3-} + e \rightarrow Co(CN)_6^{4-} \rightarrow Co(CN)_5^{3-} + CN^-.$$
$$d^6 \text{ (stable)} \qquad d^7 \text{ unstable} \qquad \text{stable} \qquad (8.2)$$

This 'free radical' species can undergo reactions of the type shown in Table 8.4. The reaction of $Co(CN)_5^{3-}$ with H_2 is a one step homolytic splitting

$$2Co^{II}(CN_5)^{3-} + H_2 \rightarrow 2Co^{III}(CN)_5H^{3-} \qquad (8.3)$$

but reactions with other molecules of the type X–Y (for example, X–Y = H_2O_2, CH_3I and other halides) take place in a stepwise fashion through the abstraction route followed by free radical addition,

$$Co(CN)_5^{3-} + X–Y \rightarrow Co(CN)_5Y^{3-} + X\cdot \text{ (rate determining)}$$
$$(8.4)$$

$$Co(CN)_5^{3-} + X\cdot \rightarrow Co(CN)_5X^{3-} \qquad (8.5)$$

TABLE 8.4

Species of related configurations and reactivities in organic and co-ordination chemistry*

| Organic species | | | | Transition metal counterpart | | | |
Species	Co-ordination number[b]	Non-bonding electrons	Characteristic reactions	Co-ordination number[b]	Non-bonding electrons	Examples	Remarks
Saturated molecule (R_3C-X)	4	0	Substitution	6	6	$RhCl_6^{3-}$	$RhCl_6^{3-} + H_2O \rightleftharpoons RhCl_5OH_2^{2-} + Cl^-$ $RhCl_6^{3-} + H_2 \rightleftharpoons RhHCl_5^{3-} + HCl$
Free radical ($R_3C\cdot$)	3	1	Dimerization Abstraction Addition	5	7	$Co(CN)_5^{3-}$	$2Co(CN)_5^{3-} \rightleftharpoons Co_2(CN)_{10}^{6-}$ $Co(CN)_5^{3-} + CH_3I \rightarrow Co(CN)_5I^{3-} + CH_3\cdot$ $2Co(CN)_5^{3-} + CH\equiv CH \rightarrow (NC)_5CoCH=CHCo(CN)_5^{6-}$
Carbene ($R_2C:$)	2	2	Addition Insertion	4	8	$IrI(CO)(PPh_3)_2$	$IrI(CO)(PPh_3)_2 + C_2H_4 \rightleftharpoons IrI(CO)(C_2H_4)(PPh_3)_2$ $IrI(CO)(PPh_3)_2 + H_2 \rightleftharpoons IrH_2I(CO)(PPh_3)_2$
Carbonium ion (R_3C^+)	3	0	Addition of nucleophile	5	6	[a] $Co(CN)_5^{2-}$	$Co(CN)_5^{2-} + I^- \rightleftharpoons Co(CN)_5I^{3-}$
Carbanion ($R_3C:^-$)	3	2	Addition of electrophile	5	8	$Mn(CO)_5^-$	$Mn(CO)_5^- + H^+ \rightleftharpoons Mn(CO)_5H$

[a] Intermediate in S_N1 substitution reactions of $Co(CN)_5OH_2^{2-}$.

[b] Note that the change in co-ordination number and in the number of nonbonding electrons in going from one species to the next is the same in each series. This results in correspondingly similar changes in the reactivity patterns of the two series of compounds since the reactivity pattern in each case is dominated by the tendency to return to the stable closed-shell configuration of the first member of the series.

* From Halpern, 1970.

H*

giving overall

$$2Co(CN)_5^{3-} + XY \rightarrow Co(CN)_5Y^{3-} + Co(CN)_5X^{3-}. \qquad (8.6)$$

Thus hydrogenolysis reactions, homogeneously catalysed by $Co(CN)_5^{3-}$, may be seen to take place according to the following reaction scheme:

$$2Co(CN)_5^{3-} + H_2 \rightleftharpoons 2Co(CN)_5H^{3-} \qquad (8.3)$$

$$2Co(CN)_5^{3-} + XY \rightarrow Co(CN)_5X^{3-} + Co(CN)_5Y^{3-} \qquad (8.6)$$

$$Co(CN)_5H^{3-} + Co(CN)_5X^{3-} \rightarrow 2Co(CN)_5^{3-} + XH \qquad (8.7)$$

$$Co(CN)_5H^{3-} + Co(CN)_5Y^{3-} \rightarrow 2Co(CN)_5^{3-} + YH \qquad (8.8)$$

overall $XY + H_2 \rightarrow XH + YH.$ $\qquad (8.9)$

There is a great tendency, as above, to return to the stable octahedral d^6 configuration (compare this with the tendency to return to closed shell configuration for organic compounds).

(d) Another important factor in homogeneous catalysis by metal complexes is the ability to promote rearrangements within co-ordination shells of two or more stable configurations of the complex differing in co-ordination number, for example 4 and 5 for d^8 complexes. Rearrangements in the OXO reaction of the type

$$CH_3Co(CO)_4 \rightleftharpoons CH_3\overset{\displaystyle O}{\overset{\displaystyle \parallel}{C}}Co(CO)_3 \qquad (8.10)$$

may be facilitated by this type of rearrangement.

Following this general survey of the applications of homogeneous catalysis, the features of the Wacker process for acetaldehyde manufacture, which is a homogeneously catalysed reaction of major importance, will be analysed in some detail. Finally it will be noted in Table 8.1 that mass transfer of oxygen into the liquid is usually the rate determining step in this type of reaction and some implications of this fact for reactor design will be discussed in the final section of the chapter.

8.3 The Wacker Process

The process is represented by the overall reaction

$$C_2H_4 + \tfrac{1}{2}O_2 \rightarrow CH_3CHO \qquad (8.11)$$

which takes place catalytically in the presence of palladium and copper chlorides in aqueous solution.

In fact ethylene and palladium chloride may be considered to react together according to

$$C_2H_4 + PdCl_2 + H_2O \rightarrow CH_3CHO + Pd + 2HCl. \qquad (8.12)$$

The palladium is then oxidised by cupric chloride according to

$$Pd + 2CuCl_2 \rightleftharpoons PdCl_2 + 2CuCl. \qquad (8.13)$$

If the cupric chloride is maintained in large excess this reaction occurs quantitatively. A small concentration of palladium salt need only be present and will therefore function as a catalyst.

The cuprous chloride is itself reoxidised in a fast reaction by oxygen or air

$$2CuCl + 2HCl + \tfrac{1}{2}O_2 \rightarrow 2CuCl_2 + H_2O. \qquad (8.14)$$

This type of oxidation is not confined solely to ethylene oxidation. Thus propylene can be oxidised to acetone, cis and trans 2-butene and 1-butene to methyl ethyl ketone while, if the reaction involving palladium chloride takes place in acetic acid containing an excess of acetate ion rather than in aqueous medium, ethylene is oxidised to vinyl acetate together with acetaldehyde.

8.3.1   KINETICS AND MECHANISM

The overall process can be considered to consist of three chemical stages which may of themselves be complex and will in fact interact with one another. It is convenient however to consider each of these separately. These three stages will be referred to as (a) ethylene oxidation stage (b) palladium oxidation stage (c) cuprous chloride oxidation stage.

8.3.1.1 *Ethylene Oxidation Stage*

The key to the Wacker process lies in the ability of palladium, which is a Group VIII transition element, to form complex ions by the co-ordination of ligands. Palladium has the capability of co-ordinating four donor groups; thus in the presence of chloride ions the complex ion $PdCl_4^{2-}$ is formed.

When ethylene is bubbled into an aqueous solution containing this complex ion, exchange between the ethylene and a chloride ion takes place according to the equilibrium equation

$$\text{(I)}$$
$$PdCl_4^{2-} + C_2H_4 \rightleftharpoons PdCl_3(C_2H_4)^- + Cl^-. \tag{8.15}$$

The double bond in the ethylene is capable of donating an electron pair to the palladium forming a π complex.

For reaction at a constant ethylene pressure of 1 atm., a plot of log [product] versus time, for reaction after the initial rapid uptake of ethylene, was found to be linear until inhibition by Cl^- ion became significant. Since the volume of ethylene consumed in excess of solubility corresponded to the amount of Pd(II) reduced, the reaction is first order with respect to Pd(II). The order in ethylene was determined by operating at different olefine pressures and was shown to be first order also. The reaction was greatly retarded by the presence of acids and, on the basis of these observations, a simplified rate equation for the formation of acetaldehyde, using a palladium complex, of the form

$$\text{rate} = \frac{k[\text{Pd } Cl_4^{2-}][C_2H_4]}{[H_3O^+][Cl^-]^2} \tag{8.16}$$

was postulated to explain the observed behaviour (Henry, 1964).

The initial complex forming reaction, (8.15), is followed by hydrolysis. The formation of an olefin-hydroxo complex, (II), through reaction of (I) with water, by a two-step process involving the elimination of a chloro ligand followed by the detachment of a hydroxonium ion, has been postulated (equations 8.17 and 8.18):

$$PdCl_3(C_2H_4)^- + H_2O \rightleftharpoons PdCl_2(C_2H_4)(H_2O) + Cl^- \tag{8.17}$$

$$PdCl_2(C_2H_4)(H_2O) + H_2O \rightleftharpoons PdCl_2(C_2H_4)(OH)^- + H_3O^+.$$
$$\text{(II)}$$
$$\tag{8.18}$$

The inverse relationship of the reaction rate with respect to the square of $[Cl^-]$ and to $[H_3O^+]$ may be inferred from consideration of equations (8.15), (8.17) and (8.18).

The next and rate determining step involves the insertion of Pd(II) and OH across the double bond to give a hydroxy palladium adduct (Henry, 1971)

The formation of acetaldehyde is then pictured as proceeding through the following transition state

$$\rightarrow CH_3CHO + Pd^0 + 2Cl^- + H_3O^+$$

(8.19)

The basis of the process is nucleophilic attack on the co-ordinated olefin, resulting from the fact that the electron density between the carbon atoms is lowered by the complexing process.

Evidence seems to be in favour of the hydrolysis step being rate determining rather than the initial complex forming step, since increasing the temperature leads to an increase in acetaldehyde formation; if the opposite were the case, the rate would decrease with increase in temperature since higher equilibrium concentrations of (I) are obtained the lower the temperature. The slowest step in the above mechanism is the rearrangement of a π-complex to a σ-complex. All four hydrogen atoms in the acetaldehyde come from the original ethylene, as shown, since no deuterium is detected in acetaldehyde when the reaction is carried out in deuterated water.

8.3.1.2 *Palladium Oxidation Stage* (Smidt *et al.*, 1962)

In the industrial process the palladium metal produced in the olefin oxidation is oxidised to the chloride. Cupric chloride is used as the oxidising agent largely because of the ease of oxidation by molecular oxygen of the cuprous chloride produced.

If there is sufficient cupric chloride present metallic palladium will not be precipitated at all, which indicates that the ethylene oxidation stage is rate determining rather than the palladium oxidation process.

Oxidation of the Pd_{met} cannot in fact take place according to

$$2Cu^{++} + Pd_{met} \rightleftharpoons 2Cu^+ + Pd^{++}, \tag{8.20}$$

since K for this reaction is $10^{-28 \cdot 2}$ at 298°K.

In the presence of chloride ions however the divalent palladium ion is stabilised as $Pd\,Cl_4{}^{2-}$ and the cuprous ion as $CuCl_2{}^-$. Hence the palladium oxidising reaction is represented by

$$2Cu^{++} + Pd_{met} + 8Cl^- \rightleftharpoons 2CuCl_2{}^- + PdCl_4{}^{2-}. \tag{8.21}$$

The equilibrium constant, K, for this reaction at 298 °K is $7 \cdot 9 \times 10^{-6}$. At high degrees of oxidation of copper, i.e. high values of

$$\frac{[Cu^{++}]}{[Cu^{++}] + [Cu^+]},$$

higher concentrations of $PdCl_4{}^{2-}$ will be favoured as deduced from consideration of the equilibrium expression for reaction (8.21)

$$K = \frac{[PdCl_4{}^{2-}][CuCl_2{}^-]^2}{[Cl^-]^8 [Cu^{++}]^2}.$$

e.g. at 298°K,   Oxidation degree $[PdCl_4{}^{2-}]$mole/l

Oxidation degree	$[PdCl_4{}^{2-}]$mole/l
0·9	0·071
0·5	$7 \cdot 9 \times 10^{-6}$

In the case where a large concentration of cupric chloride is used, the palladium chloride concentration will remain constant during the process but the hydrogen and chloride ion concentrations will change according to

$$C_2H_4 + 2Cu^{++} + 4Cl^- + H_2O \xrightarrow{PdCl_2} CH_3CHO + 2H^+ + 2CuCl_2{}^-.$$

$$\tag{8.22}$$

It has been seen, (equation 8.16), that both hydrogen and chloride ions retard the oxidation of ethylene considerably. The diminution of $[Cl^-]$ must increase the rate of reaction. The effect of Cl:Cu on rate as a function of degree of oxidation and also the change in pH are shown in Figs. 8.2 and 8.3 respectively.

In pure $CuCl_2$ (i.e. Cl:Cu = 2:1) the rate decreases immediately after starting but reducing the $[Cl^-]$, for example by replacing some of the cupric chloride with say copper oxychloride, increases the rate of reaction with decreasing $[Cl^-]$, and the decline in rate is shifted to

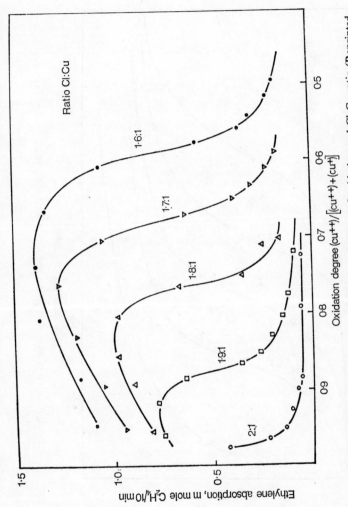

FIG. 8.2. Ethylene absorption rate as a function of degree of oxidation and Cl:Cu ratio (Reprinted with permission from *Ethylene and Its Industrial Derivatives*, Chapter 8, p. 647, Benn, London 1969)

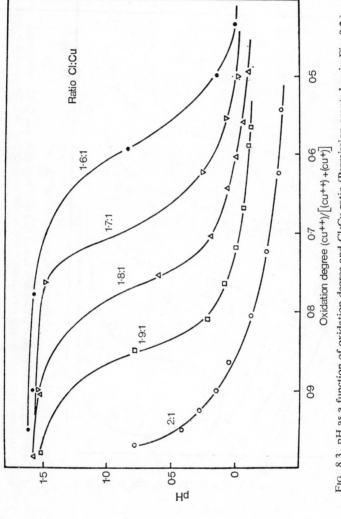

FIG. 8.3. pH as a function of oxidation degree and Cl:Cu ratio (Permission granted as in Fig. 8.2.)

lower degrees of oxidation. This increase in rate with decreasing [Cl⁻] is much greater than expected, taking account of the change in [Cl⁻] only. What happens is that the hydroxyl ions of the basic copper salt, which replace the chloride ions, are neutralised by hydrogen ions produced according to reaction (8.22). When this neutralisation is complete the rate drops steeply and it can be seen that the pH also drops simultaneously. This sudden decline occurs at lower degrees of oxidation the lower the Cl:Cu ratio.

It may be appreciated therefore that for a technical process the Cl:Cu ratio is a highly important parameter since it is desirable to maintain the rate as high as possible. It should be noted that not only is the reaction rate increased by lowering the chloride ion content, but the fast reaction range is also increased.

There are limitations on the degree to which the reaction range can be enlarged. At low [Cl⁻] precipitation of cuprous chloride and palladium metal are favoured and this can occur before neutralisation of the copper oxychloride is complete. Conversely, at high oxidation degrees a Cl:Cu lower than 2:1 leads to precipitation of a copper oxychloride of composition $Cu_2Cl(OH)_3 + xH_2O$. Hence the catalyst must not be oxidised completely.

8.3.1.3 *Cuprous Chloride Oxidation Stage*

This reaction is very fast in aqueous solution, the rate being proportional to $[CuCl_2^-]$ and $[O_2]$. The solution cannot be saturated with oxygen because of the high reaction rate, so there is a steady state in oxygen concentration.

The reaction is described by the equation

$$2CuCl_2^- + 2H^+ + \tfrac{1}{2}O_2 \rightarrow 2Cu^{++} + 4Cl^- + H_2O. \qquad (8.23)$$

8.3.2 INDUSTRIAL PROCESSES—CONDITIONS OF OPERATION

The reaction steps which have been analysed can be combined to give two variations of process (1) single stage (2) two stage.

In the single stage, as the name suggests, all steps in the reaction take place simultaneously, whereas in the two stage process the oxidation of cuprous chloride is carried out in a separate stage. Ethylene and oxygen are therefore introduced in separate parts of the plant in the two stage process.

In the single stage process a mixture of ethylene and oxygen are pumped through a vertical ceramic lined reactor charged with the catalyst solution, the reaction taking place at ≈ 3 atm. and 390–400°K. The feed contains about 9% oxygen to prevent the formation of an explosive mixture. The reaction heat evaporates the acetaldehyde produced and this and non-reacted gas are separated from the catalyst at the top of the reactor, the catalyst being recirculated to the bottom of the reactor. In this variant of the process a steady state is achieved since the reaction and oxidation occur at an equal rate. The concentrations of all catalyst components therefore remain constant and at a fixed degree of oxidation. Obviously this steady state condition should fall within the range of high reaction rates. Since low oxygen concentrations are required, relatively high Cl/Cu ratios must be used to prevent the establishment of the steady state at degrees of oxidation where precipitation of cuprous chloride could occur. In this one-stage process also, the reaction proceeds catalytically with respect to the copper in the catalyst and so a lower total concentration of copper can be used. In the stationary state both gases react stoichiometrically, i.e. in the ratio $C_2H_4 : O_2 = 2:1$, and only a small excess of oxygen is necessary to compensate for cupric chloride removed in side reactions. The oxidation degree, and hence the reaction rate, can be controlled by variation of $[Cl^-]$. As hydrochloric acid is consumed in side reactions $[Cl^-]$ is controlled by acid addition. This also prevents precipitation of copper oxychloride, cuprous chloride and Pd metal.

In the two-stage process the ethylene and oxygen are reacted in two separate tube reactors. Air is used instead of oxygen. In the first tower, in which the ethylene and palladium oxidations take place, the ethylene is fed at a pressure of 10 atm. The ethylene is 99% converted. The catalyst solution containing product is then expanded to normal pressure, the acetaldehyde and some steam flashing off. The catalyst is recycled to an oxidation unit where air re-oxidises the copper salt. The oxygen conversion is about 90%. In this variant of the process the catalyst solution is alternately reduced and oxidised. The Cl/Cu ratio must be kept within the 'fast reaction' range and this ratio must be chosen such that cuprous chloride and palladium metal do not precipitate. A high copper concentration is used because it results in a high ethylene conversion per volume of catalyst, but, as mentioned, a low $[Cl^-]$ is limited by the need to prevent precipitation of cuprous chloride and palladium metal and, at the higher degrees of oxidation, the precipitation of copper oxychloride. The redox potential for the transition Cu^{II} to Cu^{I}, according to

$$Cu^{++} + 2Cl^{-} + e \rightleftharpoons CuCl_{2}^{-}, \tag{8.24}$$

can be considered as a measure of the degree of oxidation of the solution at any time and this measurement can be used to control the process.

In both processes, because of the corrosive nature of the catalyst solution, titanium or titanium lined vessels are used.

8.4 Mass Transfer Effects in Gas-Liquid Reactions

The homogeneously catalysed reactions which have been discussed, generally involve the absorption of a gas into a liquid in which chemical reaction takes place, as, for example, in air oxidation and in hydrogenation. As can be seen from Table 8.1 mass transfer of oxygen is invariably a rate controlling step in oxidation reactions. In this section some aspects of mass transfer relevant to homogeneously catalysed reactions and to the design of reactors for such reactions will be considered. Little more than an indication of the general approach is given. Further details on gas-liquid reactions may be found in the book by Danckwerts (1970) and a discussion of reactor design problems in liquid phase oxidation is to be found in an article by Prengle and Barona (1970b).

The types of reactor employed for carrying out these reactions are sparged towers, in which agitation is produced by the bubbling of gas through a liquid, in which there may be no additional agitation, and stirred tank reactors. In the former there will be little backmixing and the bubble flow patterns may be hard to characterise, while in the latter, the degree of backmixing will be large and drop coalescence will be small. Although continuous flow of the liquid phase is most common, these reactors may also be operated on a semi-batchwise basis. The usual model assumes that gas crosses the interface between the two phases by diffusion and convective mass transfer and reacts with the liquid in a 'reaction zone' between the interface and the bulk of the liquid. Gas absorption accompanied by chemical reaction is usually studied by comparison with transfer without reaction, the difference in rate being a measure of the enhancement of mass flux of the gas brought about by reaction.

For gas absorption without reaction involving a single gas only, i.e. for the case when gas film resistance is negligible, the gas transport equation is given by

$$N_A = k_L(C_{A_i} - C_A) \qquad (8.25)$$

where N_A is average rate of transfer of gas/unit area, C_{A_i} is concentration of dissolved gas in equilibrium with the partial pressure of gas at the interface between gas and liquid, C_A is the average concentration of gas in the bulk liquid phase and k_L is the mass transfer coefficient.

Various theories have been put forward to explain mass transfer phenomena, the simplest being the film model of Whitman in which a stagnant film, of thickness δ, at the surface of the liquid next to the gas is assumed to exist. The concentration of gas across this film changes from C_{A_i} to C_A and the dissolved gas is considered to cross solely by molecular diffusion. In this model $k_L = D_A/\delta$ where D_A is the diffusion coefficient for gas A. Other models which have been presented, and in fact are probably more realistic, include the penetration theory of Higbie and the boundary layer theory. For these k_L is proportional to $(D_A)^{\frac{1}{2}}$.

Hatta's (1932) theory of mass transfer with reaction takes account of the enhancement of mass flux across the interface by reaction. When mass transfer and reaction occur simultaneously, reaction enhances this flow. For mass transfer with reaction the transfer coefficient, k_L^*, may be obtained from the expression

$$N_A^* = k_L^*(C_{A_i} - C_A) \qquad (8.26)$$

where k_L^* is calculated from

$$\phi = k_L^*/k_L, \qquad (8.27)$$

and ϕ is obtained by solution of equation (8.28) for mass transfer with chemical reaction

$$-\frac{dN_A}{dz} - (-r_A) = \frac{dC_A}{dt}. \qquad (8.28)$$

Solutions are obtained in terms of plots of ϕ vs. $\sqrt{M}$ where $M \propto k_1/k_L^2$ for the film theory, k_1 being the rate constant for the chemical reaction. If the rate and mass transport coefficients are known, Hatta's theory is satisfactory for design purposes but often this is not the case and other simpler methods are desirable.

Prengle and Barona (1970) have put forward a procedure, based on a macroscopic view of the reacting system, which is a combination of Fick's First Law and a mass balance for each component in the continuous and disperse phases. Equations have been developed for both plug flow and complete mixing of the continuous phase and also for the stirred batch case, the sparged disperse phase being either in plug flow or perfectly mixed.

Thus, for the case of a batch reactor in which the gas is sparged, the mass balance for the gas in the continuous phase is

$$a \frac{V_D}{V_C} N_A - (-r_A) = \frac{dC_A}{dt}, \qquad (8.29)$$

mass	rate of	accumu-
transfer	chemical	lation
rate	reaction	rate

where a is interfacial area between phases and V_D/V_C is (gas holdup volume)/(continuous phase volume). Combining this equation with the transport equation (8.25) gives

$$a \frac{V_D}{V_C} k_L (C_{Ai} - C_A) - (-r_A) = \frac{dC_A}{dt}. \qquad (8.30)$$

Integration of this equation gives the gas profile in the liquid, C_A/C_{Ai}, as a function of time (Fig. 8.4). In the case of batch mass transfer with reaction, the reaction cannot take place at a rate greater than the rate at which the gas can be transferred to the liquid. The maximum rate is therefore that at which $dC_A/dt = 0$

$$(-r_A)_{max} = a \frac{V_D}{V_C} k_L (C_{Ai} - C_A). \qquad (8.31)$$

In the case of continuous flow of the liquid phase, with gas bubbling through the liquid in steady flow in a stirred vessel, the relevant equation is

$$a \frac{V_D}{V_C} N_A - (-r_A) = \frac{v}{V_C} (C_A - C_{Ao}), \qquad (8.32)$$

where v is the volumetric liquid feed rate, V_C is continuous phase volume and C_{Ao} is initial gas concentration, C_A being the steady state gas concentration.

Prengle and Barona (1970) give an example based on data of Azbel (1964) for the oxidation of o-xylene to p-toluic acid at 433°K and 14 atm. according to the equation

$$\underset{\text{(B)}}{o-C_8H_{10}} + 1 \cdot 50_2 \rightarrow \underset{\text{(A)}}{o-C_7H_7COOH} + H_2O.$$

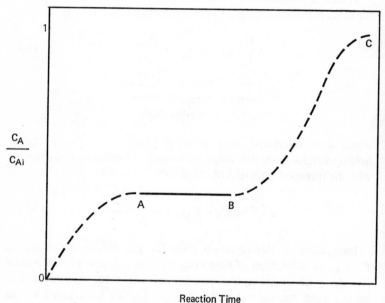

Reaction Time

FIG. 8.4. Oxygen concentration profile in a batch reactor as as function of time.

 OA—region of build up to steady state concentration

 AB—region of steady state reaction

 BC—region in which oxygen concentration builds
  up after depletion of liquid reactants

The reaction rate is given by

$$-r_A = k \, C_A{}^2.$$

The reaction takes place with oxygen being sparged into a continuous flowing stream of o-xylene. The conversion of xylene per pass is 16% and unreacted xylene is recycled to the reactor.

It is required to assess the importance of mass transfer of oxygen in the reaction from the following data:

Optimum conversion rate of oxygen, $(-r_A) = 3.34 \times 10^{-3}$ $kmol/m^3 s$.

Volumetric flow rate of o-xylene, $v = 1.543 \times 10^{-3}$ m^3/s.

Diffusivity of oxygen, $D_A = 1.497 \times 10^{-8}$ m^2/s.

Solubility of oxygen in o-xylene, $C_{A_i} = 22.92 \times 10^{-3}$ $kmol/m^3$ o-xylene. Density, $p_L = 750.3$ kg/m^3

Viscosity, $\mu_L = 2.3 \times 10^{-4}$ $N\,s/m^2$

Surface tension, $\sigma_L = 16.5 \times 10^{-3}$ N/m.

Solution. The equation to be applied is (8.32)

$$a \frac{V_D}{V_C} k_L(C_{A_i} - C_A) - (-r_A) = \frac{v}{V_C}(C_A - C_{A_0}).$$

(i) a, the interfacial area/unit volume of disperse phase, can be determined from the bubble diameter, D_p, and ψ, a shape factor

$$a = 6\psi/D_p. \tag{8.33}$$

For the case of bubbles generated at the tip of an orifice, as in the present example, D_p, may be calculated from Leibson's correlation (1956) and Hughmark's correction (1967). Details are not given here as this example is designed primarily to illustrate the principles involved. D_p for this example may be taken as 0·00289 m, and, since the bubbles are small, ψ can be taken as unity. The interfacial area, a, is then calculated to be 2078·4 m^2/m^3.

(ii) V_D/V_C.

$$\frac{V_D}{V_C} = \frac{h'}{1-h'} \tag{8.34}$$

where h' is the dynamic hold up and is related to the static hold up, h, by

$$h' = h\left[1 - \frac{u_{SL}}{u_{SG}} \frac{h'}{1-h'}\right] \tag{8.35}$$

where u_{SL} and u_{SG} are the superficial velocities of liquid and gas respectively.

h can be determined from

$$h = \left(\frac{1}{u_T}\right)\frac{\beta\, u_{SG}}{1+2\beta u_{SG}} \tag{8.36}$$

where u_T is the terminal bubble velocity and β is a correction factor for liquids other than water.

$$\beta = \left(\frac{1000}{\rho_L}\right)^{1/3}\left(\frac{0\cdot072}{\sigma_L}\right)^{1/3} \tag{8.37}$$

where ρ_L and σ_L are the density and surface tension of the liquid respectively.

Prengle and Barona show that when a value of $u_{SL} = 0\cdot00305$ m/s is used with reasonable values of u_{SG}, h and h' are of similar magnitude when evaluated according to equations (8.35) and (8.36) for this particular flow reactor. Also they show that there is an optimum value of u_{SG} which leads to minimum reactor volume. For the present case, $u_{SGopt} = 0\cdot074$ m/s and with a value of β of $1\cdot093$ from (8.37) and $\mu_T = 0\cdot265$ m/s, equation (8.36) gives for h

$$h = \left(\frac{1}{0\cdot265}\right)\left(\frac{1\cdot093\times0\cdot074}{1+2\times1\cdot093\times0\cdot074}\right) = 0\cdot264.$$

Then from equation (8.34)

$$\frac{V_D}{V_C} = \frac{h'}{1-h'} \approx \frac{h}{1-h} = 0\cdot359.$$

(iii) k_L. The mass transfer coefficient, k_L, for bubble columns is deduced from the correlation

$$Sh = 2+0\cdot0187\left[Re_p{}^{0\cdot484}Sc^{0\cdot339}\left(\frac{D_p g^{1/3}}{D_A{}^{2/3}}\right)^{0\cdot072}\right]^{1\cdot61} \tag{8.38}$$

where $Re_p = (D_p u_s \rho_L)/\mu_L$, u_s being the slip velocity of the bubble, which is given by

$$u_s = \frac{u_{SG}}{h} - \frac{u_{SL}}{1-h} = 0\cdot278 \text{ m/s for } u_{SL} = 0\cdot00305 \text{ m/s},$$

and $Sc = \dfrac{\mu_L}{\rho_L D_A}$.

Inserting numerical values in (8.38) gives

$$Sh = 2 + 0.0187[(2621)^{0.464}(20.5)^{0.339}(1018)^{0.072}]^{1.61}$$

$$= 49.7.$$

Now $Sh = \dfrac{k_L D_p}{D_A}$ \hfill (8.39)

and k_L is calculated to be 2.57×10^{-4} m/s.

(iv) V_C. V_C is the liquid phase volume required for reaction to take place.

$$V_C = \frac{v \, \rho_L}{M_L(-r_B)} \tag{8.40}$$

where M_L is molecular weight of o-xylene and $(-r_B)$ is rate of disappearance of o-xylene. Thus from equation (8.40)

$$V_C = \frac{1.543 \times 10^{-3} \times 750.3}{106.16 \times 2.23 \times 10^{-3}} = 4.895 \text{ m}^3.$$

(v) With the initial concentration of oxygen in the bulk phase $C_{A_0} = 0$, C_A may be calculated using equation (8.32):

$$(2078.4)(0.359)(2.57 \times 10^{-4})(22.92 \times 10^{-3} - C_A) - (3.34 \times 10^{-3})$$

$$= \frac{1.543 \times 10^{-3}}{4.895}(C_A - 0)$$

and $C_A = 5.51 \times 10^{-3}$ kmol/m^3.

(vi) The mass transfer contribution is therefore

$$a \frac{V_D}{V_C} k_L (C_{A_i} - C_A) = 3.34 \times 10^{-3} \text{ kmol/m}^3 \text{ s}.$$

Since $(-r_A) = 3.34 \times 10^{-3}$ kmol/m^3 s, the contribution of the $(v/V_C)(C_A - C_{A_0})$ term is very small, i.e. the rate of accumulation of oxygen is negligible and therefore the reaction is controlled by the rate of transfer of oxygen to the solution.

REFERENCES

AZBEL, D. S. 1964. *Khim. Prom.* **40**(9), 689; and **40**(12), 881.
DANCKWERTS, P. V. 1970. *Gas liquid reactions.* McGraw-Hill, London.
HALPERN, J. 1968. *Advan. in Chem. Ser.* **70**, 1.

HATTA, S. (1929) and (1932). *Tohoky Imp. Univ. Tech. Reports* **8**, 1; **10**, 119.

HENRY, P. M. 1964. *J. Amer. Chem. Soc.* **86**, 3246.

HENRY, P. M. 1971. *Trans. N.Y. Acad. Sci.* **33**, 41.

HUGHMARK, G. A. 1967. *Ind. Eng. Chem.* (*Proc. Des. & Dev.*) **6**, 218.

LEIBSON, I., HOLCOMB, E. G., CACOSO, A. G. and JACMIE, J. J. 1956. *A.I.Ch.E.J.* **2**, 296.

PRENGLE, H. W. and BARONA, N. 1970. *Hydro. Proc.* **49**(3), 106(a); **49**(11), 159(b).

SMIDT, J., HAFNER, W., JIRA, R., SIEBER, R., SEDLMEIER, J. and SABEL, A. 1962. *Angew. Chem. Int. Ed.* **1**, 80.

APPENDIX 1

Standard Heats, Free Energies and Entropies of Formation of Some Common Compounds at 298°K

(Units of $\Delta H_f°$ and $\Delta G_f°$ are kJ/mol and those of $S°$, J/mol °K; all values for gases unless stated.) (l ≡ liquid.)

Standard States:

Gases: the ideal gas state at 1 atm.
Liquid: the liquid state at 1 atm.

Carbon Compounds	$\Delta H_f°$	$\Delta G_f°$	$S°$
Methane	−74·847	−50·793	186·19
Ethane	−84·667	−32·886	229·49
Propane	−103·847	−23·489	269·91
n-Butane	−124·733	−15·707	310·12
iso-Butane	−131·595	−17·974	294·64
n-Pentane	−146·44	−8·20	348·95
	−173·05(l)	−9·247(l)	
2-Methyl butane	−154·473	−14·644	343·59
	−179·284(l)	−15·02(l)	
2-2-Dimethyl propane	−165·98	−15·23	306·39
n-Hexane	−167·19	0·21	388·40
	−198·82(l)	−3·81(l)	
n-Heptane	−187·82	8·74	427·77
	−224·39(l)	1·76(l)	
Ethylene	52·28	68·12	219·45
Propylene	20·41	62·72	266·94
1-Butene	1·17	72·04	305·60
cis-2-Butene	−5·70	66·97	300·83
trans-2-Butene	−10·06	64·11	296·48
iso-Butene	−13·99	60·98	293·59
1-Pentene	−20·92	78·60	
Acetylene	226·75	209·2	200·82
Methyl acetylene	185·43	193·77	248·11
Benzene	82·93	129·66	269·20
	49·03(l)	124·50(l)	172·80(l)
Toluene	50·00	122·29	319·74
	12·00(l)	114·15(l)	219·58(l)
Ethylbenzene	29·79	130·57	
	−12·46(l)	119·72(l)	255·18(l)
Cyclopentane	−77·24	38·62	
	−105·90(l)	36·40(l)	
Cyclohexane	−123·13	31·76	
	−156·23(l)	26·74(l)	

	$\Delta H_f{}^\circ$	$\Delta G_f{}^\circ$	S°
Methanol	$-201 \cdot 17$	$-161 \cdot 59$	$236 \cdot 31$
	$-238 \cdot 66(l)$	$-166 \cdot 52$	$126 \cdot 78(l)$
Ethanol	$-218 \cdot 53$	$-168 \cdot 32$	$282 \cdot 00$
	$-277 \cdot 61(l)$	$-174 \cdot 72(l)$	$160 \cdot 07(l)$
n-Propanol	$-255 \cdot 94$	$-162 \cdot 47$	
	$-300 \cdot 70(l)$	$-166 \cdot 69(l)$	
iso-Propanol	$-261 \cdot 12$	$-159 \cdot 83$	$307 \cdot 11$
	$-310 \cdot 95(l)$	$-162 \cdot 46(l)$	$179 \cdot 91(l)$
Ethylene glycol	$-387 \cdot 15$	$-298 \cdot 15$	
	$-451 \cdot 50(l)$	$-319 \cdot 82(l)$	
Phenol	$-90 \cdot 83$	$-26 \cdot 19$	
	$-158 \cdot 16(l)$	$-46 \cdot 11(l)$	
Ethylene oxide	$-67 \cdot 36$	$-29 \cdot 04$	
Dimethyl ether	$-180 \cdot 16$	$-109 \cdot 03$	$266 \cdot 60$
Diethyl ether	$-272 \cdot 80(l)$	$-116 \cdot 11(l)$	
Acetaldehyde	$-166 \cdot 19$	$-131 \cdot 63$	
Propionaldehyde	$-205 \cdot 64$	$-142 \cdot 09$	
Benzaldehyde	$-40 \cdot 04$	$24 \cdot 48$	
	$-88 \cdot 83(l)$	$9 \cdot 37(l)$	
Acetone	$-216 \cdot 69$	$-152 \cdot 51$	$304 \cdot 18$
	$-248 \cdot 19(l)$	$-155 \cdot 48(l)$	$200 \cdot 41(l)$
Acetic acid	$-438 \cdot 15$	$-342 \cdot 67$	
	$-486 \cdot 18(l)$	$-391 \cdot 45(l)$	$159 \cdot 83(l)$
Propionic acid	$-455 \cdot 01$	$-369 \cdot 32$	
	$-509 \cdot 19(l)$	$-38 \cdot 46(l)$	
Ethyl acetate	$-426 \cdot 85$	$-313 \cdot 51$	
	$-463 \cdot 25(l)$	$-318 \cdot 44(l)$	
Methylamine	$-28 \cdot 03$	$27 \cdot 61$	$241 \cdot 54$
Ethylamine	$-51 \cdot 21$	$41 \cdot 88$	
Styrene		$213 \cdot 80$	$345 \cdot 10$
			$237 \cdot 57(l)$

Inorganic Compounds	$\Delta H_f{}^\circ$	$\Delta G_f{}^\circ$	S°
Water	−241·826	−228·593	188·72
	−285·840(l)	−237·191(l)	69·94(l)
Ammonia	−45·86	−16·33	192·51
Hydrogen chloride	−92·31	−95·30	186·68
Carbon monoxide	−110·52	−137·269	
Carbon dioxide	−393·514	−394·384	
Carbon disulphide	117·61	67·49	
Nitric oxide	90·37	86·69	
Nitrogen dioxide	33·30	51·30	
Ozone	141·75	162·59	237·65
Sulphur dioxide	−296·81	−299·91	248·53
Hydrogen sulphide	−19·96	−32·84	205·64
Sulphur trioxide	−394·93	−370·66	256·23
Hydrogen	0	0	130·59
Oxygen	0	0	205·03
Nitrogen	0	0	191·49

Values in the main selected from

(i) Selected Values of Physical and Thermodynamic Properties of Hydrocarbons and Related Compounds. API Research Project 44, Carnegie Institute of Technology, Pittsburgh, 1953 and later supplements.

(ii) Selected Values of Chemical Thermodynamic Properties. National Bureau of Standards, Circular 500, 1952.

Molar Heat Capacity of Gases in Ideal Gas State
(298–1500°K)

Constants for the equation $C_p° = \alpha + \beta T + \gamma T^2$ where T is in °K, and $C_p°$ in J/mol°K

Compound	α	$\beta \times 10^3$	$\gamma \times 10^6$
Methane	14·146	75·496	−17·981
Ethane	9·401	159·832	−46·229
Propane	10·083	239·304	−73·358
n-Butane	16·083	306·896	−94·788
n-Pentane	20·481	377·033	−117·315
n-Hexane	25·150	446·458	−139·591
n-Heptane	29·681	516·502	−162·000
Ethylene	11·841	119·666	−36·510
Propylene	13·611	188·765	−57·489
1-Butene	16·355	262·956	−82·077
1-Pentene	22·372	230·494	−103·483
Acetylene	30·673	52·810	−16·272
Benzene	−1·711	324·766	−110·579
Toluene	2·410	391·175	−130·654
Methanol	18·384	101·562	−28·681
Ethanol	29·246	166·276	−49·898
Acetaldehyde	14·075	149·461	−51·195
Carbon monoxide	26·861	6·966	−0·820
Carbon dioxide	25·999	43·497	−14·832
Hydrogen	29·066	−0·837	2·012
Oxygen	25·723	12·979	−3·862
Nitrogen	27·296	5·230	−0·004
Sulphur dioxide	29·773	39·798	14·690
Sulphur trioxide	25·426	98·479	−2·874
Water	30·359	9·615	1·184
Ammonia	25·464	36·869	−6·301
Hydrogen chloride	28·167	1·812	1·548
Hydrogen sulphide	27·874	21·481	−3·573

APPENDIX 2

Free Energy Function for the Ideal Gas State at 1 atm. and Standard Heat of Formation at 0°K.

$$-(G_T - H_0°)/T, \text{ J/mol°K}$$

	Temperature, °K							$\Delta H_f°_0$ kJ/mol
	298	400	500	600	800	1000	1500	
Methane	152·55	162·59	170·50	177·36	189·16	199·37	221·08	−66·89
Ethane	189·41	201·84	212·42	222·08	239·70	255·68	290·62	−69·11
Propane	220·62	236·31	250·24	263·30	287·61	310·03	359·24	−81·51
n-Butane	244·93	265·73	284·13	301·29	333·17	362·33	426·56	−99·03
iso-Butane	234·64	254·05	271·75	288·49	319·87	348·86	412·71	−105·86
n-Pentane	269·95	295·26	317·73	338·74	377·86	413·67	492·54	−113·93
n-Hexane	295·47	325·31	351·92	376·81	423·17	465·72	559·15	−129·33
Ethylene	184·01	195·02	203·93	212·13	226·73	239·70	267·52	60·76
Propylene	221·54	235·94	248·19	259·62	280·49	299·45	340·70	35·43
1-Butene	247·90	266·27	282·50	297·65	325·60	351·16	406·98	20·75
iso-Butene	236·27	254·80	271·00	286·27	314·43	340·12	396·06	4·10
Graphite(s)	2·164	3·448	4·795	6·180	8·945	11·594	17·493	0
Hydrogen	102·19	110·55	116·94	122·18	130·48	136·98	148·91	0
Water	155·50	165·29	172·77	178·93	188·82	196·69	211·70	−238·94
Carbon monoxide	168·82	177·37	183·87	189·21	197·71	204·43	215·74	−113·81
Carbon dioxide	182·23	191·74	199·44	206·01	217·13	226·39	244·68	−393·16
Oxygen	175·98	184·56	191·10	196·51	205·20	212·12	225·13	0
Nitrogen	162·41	170·96	177·46	182·79	191·25	197·93	210·39	0
Acetylene	167·26	177·61	186·22	193·77	206·69	217·59	239·45	227·31

Enthalpy above 0°K *for Ideal Gas State at* 1 *atm.*

$(H_T° - H_0°)$, kJ/mol

	Temperature, °K						
	298	400	500	600	800	1000	1500
Methane	10·029	13·903	18·263	23·217	34·815	48·367	88·408
Ethane	11·949	17·974	25·146	33·539	53·388	76·483	144·348
Propane	14·694	23·246	33·639	45·731	74·308	107·403	203·552
n-Butane	19·435	30·710	44·329	60·149	97·337	140·331	264·722
iso-Butane	17·891	29·137	42·886	58·869	96·274	139·369	263·801
n-Pentane	23·552	37·455	54·266	73·755	119·528	172·339	324·783
n-Hexane	27·706	44·267	64·266	87·412	141·754	204·388	384·970
Ethylene	10·565	15·527	21·409	28·167	43·848	61·756	113·386
Propylene	13·544	20·878	29·606	39·714	63·388	90·751	169·745
1-Butene	17·205	27·129	39·120	52·928	85·228	122·382	229·450
iso-Butene	17·079	27·288	39·388	53·346	85·730	122·884	230·12
Graphite	1·0525	2·1037	3·4351	5·0133	8·7094	12·866	24·326
Hydrogen	8·468	11·426	14·349	17·278	23·168	29·145	44·745
Water	9·906	13·364	16·843	20·427	27·989	36·016	57·940
Carbon monoxide	8·672	11·647	14·602	17·613	23·849	30·363	47·522
Carbon dioxide	9·364	13·367	17·669	22·270	32·173	42·769	71·145
Nitrogen	8·6705	11·6415	14·581	17·564	23·717	30·135	47·085
Oxygen	8·661	11·681	14·744	17·903	24·501	31·367	49·270
Acetylene	10·006	14·816	20·046	25·635	37·652	50·585	85·943

Data from: Selected Values of Physical and Thermodynamic Properties of Hydrocarbons and Related Compounds. API Research Project 44, Carnegie Inst. of Tech., Pittsburgh, 1953 and later supplements.

INDEX

Acetaldehyde synthesis, Wacker process, 207, 216
Acetylene synthesis, 154–68
  BASF reactor, 161–2
  composition of exit gases, 163
  description, 162–3
  operating conditions, 163
  concentration of acetylene and ethylene, dependence of residence time, 158–9
  electric arc process, 165
  Hüls electric arc reactor, 165
  description, 166
  reaction conditions, 166
  utility requirements, 166
  kinetics of formation from methane, 156–8
  partial oxidation process, 161
  kinetics of, 163
  product distribution, effect of residence time and temperature, 164
  standard free energy and enthalpy changes, 155–6
  thermal pyrolysis process, 166
  thermodynamics of synthesis from hydrocarbons, 154
  Wulff thermal reactor, 166–7
  description, 166
  reaction conditions, 166, 168
Activity, 7–9, 18, 28–30
Adiabatic operation, 45, 63, 89, 114, 116
American Petroleum Institute, 9
Ammonia synthesis, 22, 57, 72–92, 94, 113, 153, 185
  catalysts, 83, 91–2
  catalyst promoters, 92
  equilibrium constant, effect of temperature, 73
  equilibrium yield, effect of temperature and pressure, 72, 74
  fixed bed catalytic reactor, 75
  heat of reaction, 72, 73, 184
  mechanism, 91–2
  operating conditions, 83
  pressure drop in catalyst bed, 79
  rate of reaction, 74
  variation with temperature and conversion, 80
  reactor output, 78
  reactor temperature profiles, 86
  reactor types, 83
  interbed cooled reactors, 86–9
  tube cooled reactors, 84–6, 89
  reactors, heat exchange in, 89
  reactors, optimum temperature and concentration gradients, 82, 89
  space time yield, 78, 83
  effect of space velocity, 79
  specific rate of formation, 78–80
  synthesis loop, 89–91
  temperature optimisation of rate, 81, 83
  yield, effect of space velocity, 76–80
Axial conduction, 53
Axial diffusion, 51, 52

Backmix reactor (*see* Continuous stirred tank reactor)
BASF reactor, 161–2
Batch reactor, 60, 158, 168, 170, 175, 176, 179, 180, 227
  rate equations for, 31–2
Benzene chlorination (*see* Chlorination of benzene)
Boudouard reaction, 108
Boundary layer theory, 226
Bubble column, 230

Calcium carbonate decomposition, 29
Calorific value, 104
Carbon formation, 98, 108, 110
Catalyst, 83, 91, 92, 94, 111, 112, 116, 117, 187, 207–10, 224
Catalyst bed, pressure drop in, 79, 188
Catalyst promoter, 92
Chain reaction, 124
  initiation, 124; propagation, 125; termination, 125

239

Change of state, irreversible, iso-
thermal, 2, 3
Chemical potential, 5, 6, 8
    effect of pressure on, 6; standard
        state, 7
Chlorination of benzene, 175–80
    conditions for industrial operation,
        179–80
    kinetics in batch reactor, 175–6
    kinetics in continuous stirred tank
        reactor, 175–8
    product distribution, 178–9
Coke formation, 136–7
σ Complex, 219
π Complex, 218–9
Complex ion, 217–8
Conduction, axial, 53
Conductivity, effective, 50, 52, 196–7
Consecutive reactions, 153
    kinetics, first order, 37, 153, 157
    maximum concentration of product
        in, 38, 158–9
Constant volume reactor, 32
Continuous phase, 227
Continuous stirred tank reactor, 40,
    168, 173, 175, 179–80, 190–1
    design equations, 42
        first order irreversible reaction, 43,
            180
        first order reversible reaction, 44
        parallel-consecutive reactions, 173–4
Continuous stirred tank reactors in
    series, 44
Co-ordination compounds of transi-
    tion metals, use as catalysts, 207
Cracking of hydrocarbons (*see* Hydro-
    carbon cracking)
Criterion of equilibrium, 1, 3
Cuprous chloride oxidation, Wacker
    process, 223

Data
    enthalpy functions, 238
    free energy functions, 237
    molar heat capacity of gases, 236
    standard heats, free energies and
        entropies of formation, 233–5
Design equations
    continuous stirred tank reactor, 42–4
    tubular plug flow reactor, 40–2
        adiabatic, 45–7
        non-adiabatic,   non-isothermal,
            47–56.

Design of pyrolysis reactors
    simplified approach to, 140–5
    use of kinetic models, 139
Diffusion, axial, 51, 52, 196
Diffusivity, effective, 49, 51, 196
Disperse phase, 227
Driving force, 1, 4

Effective conductivity, 50, 52, 196–7
Effective diffusivity, 49, 51, 196
Electric arc process, acetylene syn-
    thesis, 165
Electron configuration of transition
    metal atoms, 211
Electron distribution in transition
    metal ions, 212
Endothermic reaction, 12, 21, 65, 107,
    122, 156, 212
Energy balance equation, 46, 47, 52,
    196–7
Enthalpy change, 3
Enthalpy function, 10, 238
Entropy, absolute, 15
    change, 15
        in surroundings, 2; in system, 2;
            total, 3
        standard, 9, 15
Equilibrium approach, 95
    factors affecting magnitude, 134–5
    in hydrocarbon cracking reactions,
        133, 137
Equilibrium composition, 9, 17, 23, 27
    computation of, 22
    in steam reforming of methane, 95–
        100
    in steam reforming of higher hydro-
        carbons, 101–3
Equilibrium constant, 5, 7–8, 14, 17–9,
    21, 26, 35, 58, 67, 95, 96, 110, 133,
    220
    effect of temperature on, 12, 58, 59,
        73, 96, 110
Equilibrium conversion, 60, 64
    effect of parameter variation on,
        21–2
Equilibrium, criterion of, 1, 3
Equilibrium process, 3
Equilibrium thermodynamics, 1, 31,
    124
Equilibrium yield, 1, 77, 78, 83, 91
    effect of temperature on, 57

Ethane decomposition
  factors affecting equilibrium approach, 133
  kinetics, 127–8
Ethane reactions, standard heat and free energy changes, 122–3
Ethylene hydration, 23
Ethylene oxidation, Wacker process, 217
Ethylene synthesis, 121, 158–9
Exothermic reaction, 13, 21, 46, 57, 66, 68, 73, 80, 107, 114, 161, 182, 194
Expansion factor, 140, 142, 150

Falling temperature sequence, 68, 80, 83
Fauser–Montecatini reactor, 87, 89
Feasibility of reaction, 1, 4, 5, 8, 17
Fick's first law, 49, 227
Finite difference equations, 53–5, 197–8
First law of thermodynamics, 2, 6
First order reaction, 32, 33, 36, 37, 41, 57, 59, 140, 157, 218
Fixed bed reactor, 75, 82, 187, 188, 194, 204–5
Flow reactors (see Continuous stirred tank and Plug flow reactors)
Fluidised bed reactor, 187, 188, 191
Free energy change, 3, 4, 5, 8
  standard, 4, 17, 19, 123
Free energy data, standard, 9
Free energy function, 9, 10, 237
Free energy of formation, 4, 9
  standard, 4, 9–11, 233–5
  standard, low carbon number hydrocarbons, 155
Free energy of reaction, standard 10, 14
  effect of temperature, 12
Free energy, partial molar, 5
Free radical, 125–30, 214, 215
Fuel gas, 94, 97–8, 107
Fugacity, 4, 7, 9, 18, 19, 21, 28, 29, 75
Fugacity coefficient, 18, 19, 29, 74
  generalised chart for, 19–20

Gas absorption, 225
  with reaction, 225
Gas composition
  in steam reforming of methane, 97–100
    in steam reforming of naphtha, 104–6

in steam reforming of saturated hydrocarbons, 101–3
leaving primary reformer, 114; leaving secondary reformer, 114; leaving shift converter, 117
Gas, ideal, 19, 21
  inert, 22, 156, 190
Gas-liquid reactions, 225
  mass transfer effects in, 225
  mass transfer of oxygen, 225, 229
  types of reactor for, 225
Gas mixture, ideal, 18, 19, 21
Gas phase reactions,
  homogeneous, multiple, 26
  homogeneous, single, 18

Hatta, theory of mass transfer with reaction, 226
Heat capacities, molar, standard, 13–4, 236
Heat change, 2–3, 122–3
  in surroundings, 2, 3; in system, 2, 3
Heat exchange, 89
Heat flux in hydrocarbon cracking reactions, 134–6
Heat of cracking, 142, 144
Heat of formation, standard, 9–14, 233–5
Heat of reaction, 11, 72, 73, 107, 184,
  exothermic catalytic reactions (standard), 182
Heat transfer coefficient
  wall, 48, 50, 196; overall, 48, 198
Heat transfer in packed beds, 49, 50
Heterogeneous reactions, 29, 129
  catalysts for, 207
Heterolytic splitting, 214
Hold up, dynamic and static, 229
Homogeneous catalysis, 207–16
  processes utilising, 208–10
Homolytic splitting, 214
Hüls electric arc, 165–6
Hybridisation, 211, 213
Hydration of ethylene, 23
Hydrocarbon chlorination reactions, 175 (see also Chlorination of benzene)
Hydrocarbon cracking, 121–51, 153
  chain reactions in, 124
  ethane decomposition, kinetics, 127–8
  ethane reactions, standard heat and free energy changes, 122–3
  ethylene formation, 121

Hydrocarbon cracking (*Continued*)
  isomerisation reactions in, 129–30
  kinetics, 124–33
  kinetic models for, 137–9
  naphtha cracking, effect of reaction
    conditions, 133
  propane decomposition kinetics, 126
    product distribution, effect of
      conversion, 131
  propane reactions, standard heat
    and free energy changes, 122,
    123
  propylene formation, 121
  reactors for (*see* Pyrolysis reactors)
  secondary reactions in, 122, 130, 133,
    138
  self-inhibition, 128–9
  steam dilution, 124, 134–5, 137
  surface reactions, 129
  thermodynamics, 121–4
Hydrogen production, 94, 113
Hydrogenolysis, 216

Industrial processes
  acetaldehyde synthesis, Wacker pro-
    cess, 207, 216–25
  acetylene manufacture, 154–68
  ammonia synthesis, 72–92
  hydrocarbon cracking, 121–51
  phthalic anhydride synthesis, 182–
    205
  steam reforming of hydrocarbons,
    94–119
Inert gas, 22, 156, 190
Interfacial area, 227, 229
Internal energy change, 2
Irreversible process, 2, 3
Irreversible reaction, 65
  kinetics, first order, 32, 140, 218
  kinetics, second order, 33, 35, 180
Isomerisation in hydrocarbon cracking, 129–30

JANAF Tables, 10

Kellogg quench converter, 88–9
Kinetic models
  for pyrolysis reactors, 137–9, 139–45
  one dimensional for reaction in a
    fixed bed, 196, 198–202
  two dimensional for reaction in a
    fixed bed, 196–8, 202–5

Kinetics, 31, 111, 124, 132, 156, 163,
  170, 173, 185, 189, 194, 217

Lewis–Randall rule, 18
Ligand, 211, 213, 217, 218
Ligand field theory, 213
Liquid phase oxidation, 225
Liquid phase reaction, equilibrium, 28

Mass balance equations, 45, 47, 50,
  196–7, 227
Mass transfer coefficient, 230
Mass transfer effects in gas liquid
  reactions, 225
Mass transfer in packed beds, 49
Mass transfer of oxygen, 225, 229
Mass transfer theories
  boundary layer 226; Hatta 226;
    Higbie, 226; Whitman film, 226
Mass transfer with reaction, 227
Maximum concentration of inter-
    mediate product
  consecutive reactions, 38, 158–9
  parallel-consecutive reactions, 173–4
Maximum reaction rate, first order
  reversible, 68–72
Mean residence time, 40
Metal complexes, electron configura-
    tion, 213
  shape, 211
Methane as feedstock for acetylene
  synthesis, 154
Methane chlorination, 175
Methane-steam reaction, 27, 94, 107,
  111 (*see* also Steam reforming of
  hydrocarbons)
Methanol synthesis, 22, 57, 94
Minimum reactor volume, 67
Minimum steam ratio, 108–11
Multiple gas phase reaction, 26

Naphtha, 94, 101, 107, 154, 166
Naphtha cracking, effect of reaction
  conditions, 133
Naphthalene oxidation (*see* Phthalic
  anhydride synthesis)
Naphthoquinone oxidation, 185
National Bureau of Standards, 9

Olefine manufacture, 121
Olefine-hydroxo complex, 218

Optimisation of reaction rates, reversible exothermic reaction, 66–72
Optimum temperature sequence, 68, 70
Orbitals, 211
Order of reaction, 31, 111, 112, 131, 140
Oxidation degree, 220
Oxidation, liquid phase, 225
Oxo reaction, 94, 216

Palladium oxidation, Wacker process, 219
Parallel-consecutive reactions, 168, 169
  in a batch or tubular reactor
    kinetics 170–3; maximum concentration of intermediate product, 173
  in a continuous stirred tank reactor
    kinetics, 173–4; maximum concentration of intermediate product, 174
Parallel reactions, 153
Parametric sensitivity, 201
Partial molar free energy, 6
Partial molar volume, 7
Partial oxidation process for acetylene synthesis, 161–4
Peclet number, 49, 191
Phase change, reversible, 15
Phthalic anhydride synthesis, 182–205
  catalysts, 'American' and 'German', 187
  heat evolution in process, 186–7
  kinetics, general, 185–6
  in fixed bed reactor, 188
    effects of variables on conversion, 198, 204–5
    kinetics, one dimensional model, 194–6, 198–202
    kinetics, two dimensional model, 196, 202–5
  in fluidised bed reactor, 187, 188
    factors influencing choice of operating conditions, 189
    kinetics, 189–91
    product distribution, effect of residence time and temperature, 191–3
  mechanism of process, 185–6
  product concentrations, effect of residence time and temperature, 191–3

rate constants, 186, 196
reaction scheme, 185–6
thermodynamics of process, 183
  standard free energy and heat of reaction, 184
Plug flow reactor
  design equations, 40–1, 60, 67, 140–2
    adiabatic, 45–7
    first order irreversible, 41–2
    first order reversible, 42, 60, 63
    non-isothermal, non adiabatic, 47–56, 196
Pressure drop factor, 142, 144, 147, 149
Pressure drop in catalyst beds, 79, 188
Pressure drop in reactors for hydrocarbon cracking, 134, 142, 14
Pressure gradients in pyrolysis reactors, 140
Primary reformer, 113
Process, irreversible, 2; isothermal, 3; reversible, 1, 2; spontaneous, 2
Processes involving homogeneous catalysts, 208–10
Propane decomposition
  factors affecting equilibrium approach, 134
  kinetics, 126
  product distribution as function of conversion, 131
Propane dehydrogenation, 11, 15
Propane reactions, standard heat and free energy changes, 122–3
Pyrolysis of hydrocarbons (see Hydrocarbon cracking)
Pyrolysis reactors
  conditions of operation, 133–7
  description of, 136
  design
    simplified approach to design, 140–5
    use of kinetic models, 139
  design example, propane pyrolysis, 145–51

Radial temperature gradients, 49, 194, 203
Rate constant, 31, 75, 186
  effect of temperature on, 62, 146
Rate constant ratio, 175
Rate expressions, differential, 31–5, 37, 45, 140, 170, 189, 194
Rate of reaction, 31, 60, 74, 114, 140, 195, 218, 228

Reaction rate equations, batch
  consecutive, first order, 37–9
  irreversible, first order, 32
  irreversible, second order, 33, 35
  reversible, first order, 36, 60, 67
Reactors (*see* also Batch, continuous
    stirred tank and plug flow)
  acetaldehyde, 223–5
  acetylene from hydrocarbons, 161,
      165, 166
  ammonia synthesis, 83–9
  hydrocarbon pyrolysis, 136
  phthalic anhydride, 188
Reduced pressure, 19
Reduced temperature, 19
Refinery off gases, 154
Reformer, primary, 113
Reformer, secondary, 114
Reforming (*see* Steam reforming of
    hydrocarbons)
Reversible phase change, 15
Reversible process, 1–3
Reversible reactions, 35, 74
  exothermic, 13, 44, 57, 72, 80, 114
    effect of reaction time on yield,
        60–5
    first order, effect of temperature
        on equilibrium yield, 57–60
    optimisation of reaction rates,
        66–72
Rice and Herzfeld, 125, 127, 137, 138

Secondary reactions, 122, 130, 133, 138
Secondary reforming, 113–4
Second order reactions
  irreversible, constant volume, 33, 180
  irreversible, variable volume, 35
Shift reaction (*see* Water–gas shift
    reaction)
Space time, 41, 131
Space time yield, 78–9, 83
Space velocity, 41, 65, 76–80, 83
Spontaneous process, 2–4
Steam dilution, effect in cracking
    reactions, 124, 134–5, 137
Steam reforming of hydrocarbons,
    94–119
  equilibrium constants for carbon
      forming reactions, 110
  equilibrium constants for methane–
      steam reaction, 96
  equilibrium constants for water–gas
      shift reaction, 96

formation of carbon in, 98, 108–10
gas composition in hydrocarbon
    reforming, effect of variables,
    101–3
gas composition in methane reform-
    ing, effect of variables, 97–100
gas composition in naphtha re-
    forming, effect of variables,
    104–6
heat of reaction, 107
kinetics, 111–3
mechanism of methane reforming
    reaction, 111
mechanism of reforming of higher
    hydrocarbons, 112
reaction rate equations, 111–2
steam–hydrocarbon ratios, 108, 109,
    111
thermodynamics of methane re-
    forming, 95–100
thermodynamics of reforming of
    higher hydrocarbons, 101
Sulphur dioxide oxidation, 57
  standard heat of reaction, 187
Surface reactions, 129
Surroundings, 2
Synthesis gas, 94, 97, 104, 105
  preparation, 113
Synthesis loop, ammonia, 89–91
System, 2

Temkin equation, 74, 78, 80, 82
Temperature profiles, reactor, 86, 89,
    140
Temperature sequence, falling, 68, 80,
    82–3
  optimum, 70
Thermal pyrolysis process for acety-
    lene synthesis, 166
Thermodynamic data (*see* Data)
Third law of thermodynamics, 11, 15
Towns gas, 104, 106
Transition metal complexes
  ammines, 211, 212
  factors governing importance in
      homogeneous catalysis, 213
  reactivities of, 215
  shape of, 211
Transition metals, co-ordination com-
    pounds in homogeneous catalysis,
    207
Transition metals, electron configura-
    tion, 212

Transition metal ions, electron distribution in, 212
Tubular reactor, 40, 45, 67, 82, 140, 159, 168, 170, 189, 195
(*see* also Plug flow reactor)

van't Hoff equation, 12, 57
Volume of reactor, minimum, 67
Volume of system, variation with conversion, 33

Wacker process for acetaldehyde manufacture, 207, 216–25
  conditions of operation, 223
  cuprous chloride oxidation stage, 223
  ethylene oxidation stage, 217–9
  industrial processes
    single stage, description, 223
    two stage, description, 224
  kinetics and mechanism, 217–23
  palladium oxidation stage, 219–23

Wall heat transfer coefficient, 48, 50, 196
Water gas shift reaction, 94, 107, 111, 114–9
  catalyst for, 116–7
  conditions of operation, 116, 117
  equilibrium constants for, 96
  steam requirement, 115, 116
Whitman film theory for mass transfer, 226
Work, 2
Wulff process for acetylene formation, 166–7

Yield, 1, 21, 57, 77, 78, 83
  ammonia synthesis, effect of space velocity, 76–80
  ammonia synthesis, equilibrium, 72, 74
  effect of reaction time in a reversible exothermic reaction, 60
  space time, ammonia synthesis, 78, 79, 83